T0211596

Lecture Notes in Artificial Intelligence   10180

Subseries of Lecture Notes in Computer Science

More information about this series at http://www.springer.com/series/1244

Paolo Ciancarini · Francesco Poggi
Matthew Horridge · Jun Zhao
Tudor Groza · Mari Carmen Suarez-Figueroa
Mathieu d'Aquin · Valentina Presutti (Eds.)

Knowledge Engineering and Knowledge Management

EKAW 2016 Satellite Events, EKM and Drift-an-LOD
Bologna, Italy, November 19–23, 2016
Revised Selected Papers

 Springer

Editors

Paolo Ciancarini
University of Bologna
Bologna
Italy

Tudor Groza
Garvan Institute of Medical Research
Darlinghurst, NSW
Australia

Francesco Poggi
University of Bologna
Bologna
Italy

Mari Carmen Suarez-Figueroa
Universidad Politecnica de Madrid
Boadilla del Monte
Spain

Matthew Horridge
Stanford University
Palo Alto, CA
USA

Mathieu d'Aquin
The Open University
Milton Keynes
UK

Jun Zhao
Lancaster University
Lancaster
UK

Valentina Presutti
STLab ISTC-CNR
Rome
Italy

ISSN 0302-9743   ISSN 1611-3349   (electronic)
Lecture Notes in Artificial Intelligence
ISBN 978-3-319-58693-9   ISBN 978-3-319-58694-6   (eBook)
DOI 10.1007/978-3-319-58694-6

Library of Congress Control Number: 2017940247

LNCS Sublibrary: SL7 – Artificial Intelligence

Printed on acid-free paper

This Springer imprint is published by Springer Nature
The registered company is Springer International Publishing AG
The registered company address is: Gewerbestrasse 11, 6330 Cham, Switzerland

Preface

This volume contains the proceedings of the Satellite Events of the 20th International Conference on Knowledge Engineering and Knowledge Management (EKAW 2016), held in Bologna, Italy, during November 19–23, 2016. This edition of the conference was concerned with the impact of time and space on the representation of knowledge. Knowledge engineering has mostly been about creating static, universal representations. Yet the world is rarely static: Everything changes, including the models, and real-world systems need to evolve along with the surrounding world. Moreover, what makes some representations valid in some contexts may make them invalid elsewhere (e.g., jurisdiction for laws).

The special focus of this year's EKAW was "evolving knowledge," which concerns all aspects of the management and acquisition of knowledge representations of evolving, contextual, and local models. This includes change management, trend detection, model evolution, streaming data and stream reasoning, event processing, time- and space-dependent models, contextual and local knowledge representations, etc.

EKAW 2016 also put a special emphasis on the evolvability and localization of knowledge and the correct usage of these limits.

In addition to this specific focus, EKAW as usual covered all aspects of eliciting, acquiring, modeling, and managing knowledge, the construction of knowledge-intensive systems and services for the Semantic Web, knowledge management, e-business, natural language processing, intelligent information integration, personal digital assistance systems, and a variety of other related topics.

For the main conference we invited submissions for research papers that presented novel methods, techniques, or analysis with appropriate empirical or other types of evaluation, as well as in-use papers describing applications of knowledge management and engineering in real environments. We also invited submissions of position papers describing novel and innovative ideas that were still at an early stage. Papers accepted for the main conference are published as regular research papers in the EKAW 2016 Springer conference proceedings (LNAI 10024). Position papers are included in this volume.

In these proceedings we have gathered papers related to the other satellite events at EKAW 2016.

The main event hosted the following satellite workshops:

- OWLED-ORE 2016 — 13th OWL: Experiences and Directions Workshop and 5th OWL Reasoner Evaluation Workshop
- EKM – Second International Workshop on Educational Knowledge Management
- Drift-a-LOD – First Workshop: Detection, Representation and Management of Concept Drift in Linked Open Data
- LK&SW-2016 – Third Workshop on Legal Knowledge and the Semantic Web.

The organizers of the EKM and Drift-a-LOD workshops have provided summaries. They also selected the best papers of their workshops whose carefully revised versions are included in these proceedings.

Moreover, there were two tutorials:

- Mapping Management and Expressive Ontologies in Ontology-Based Data Access, by Diego Calvanese, Benjamin Cogrel, and Guohui Xiao, and
- Modeling, Generating and Publishing Knowledge as Linked Data, by Anastasia Dimou, Pieter Heyvaert, and Ruben Verborgh (included in this volume).

The workshop and tutorial programs were chaired by Matthew Horridge from Stanford University, USA, and Jun Zhao from the University of Lancaster, UK.

This volume also contains the accepted contributions for the EKAW 2016 demo and poster sessions. We asked for contributions that were likely to stimulate critical or lively discussions about any of the areas of the EKAW conference series. We also invited developers to showcase their systems and the benefit they can bring to a particular application. The demo and poster program of EKAW 2016 was chaired by Tudor Groza from the Garvan Institute of Medical Research, Australia, and Mari Carmen Suarez-Figueroa of the Universidad Politecnica de Madrid, Spain.

Last but not least, the EKAW 2016 program included a doctoral consortium that provided PhD students working on the topics of the conference with an opportunity to present their research ideas and results in a stimulating environment, to get feedback from mentors who are experienced research scientists in the community, to explore issues related to academic and research careers, and to build relationships with other PhD students from around the world. The doctoral consortium was intended for students at each stage of their PhD. The presenters had an opportunity to present their work to an international audience, to be paired with a mentor, and to discuss their work with experienced scientists from the research community. All accepted papers are included in this volume. The doctoral consortium was organized by Mathieu d'Aquin from the Open University, UK, and Valentina Presutti from ISTC-CNR in Italy.

The conference organization staff also included the program chairs, Fabio Vitali from the University of Bologna, Italy, and Eva Blomqvist from Linköping University, Sweden. Silvio Peroni, University of Bologna, Italy, was the sponsorship chair. Angelo Di Iorio and Silvio Peroni from the University of Bologna, Italy, took care of local arrangements. Andrea Giovanni Nuzzolese, ISTC-CNR, Italy, acted as Web presence chair, and Francesco Poggi, University of Bologna, Italy, acted as proceedings chair. Paolo Ciancarini, University of Bologna, Italy, was the general chair of EKAW 2016.

We want to thank EasyChair and Springer for their excellent cooperation regarding, respectively, supporting our conference organization and the publication of the proceedings.

Thanks to everybody, including attendees at the conference, for making EKAW 2016 a successful event.

March 2017

Paolo Ciancarini
Francesco Poggi
Matthew Horridge
Jun Zhao
Tudor Groza
Mari Carmen Suarez-Figueroa
Mathieu d'Aquin
Valentina Presutti

Organization

Program Committee

Doctoral Symposium

Alessandro Adamou	Knowledge Media Institute, The Open University
Silvio Cardoso	Luxembourg Institute of Science and Technology, Luxembourg
Marilena Daquino	University of Bologna, Italy
Median Hilal	Johannes Kepler University, Austria
David Noël	Laboratoire d'Informatique de Grenoble, France
Irlan Grangel	University of Bonn, Germany
Matthias Jurisch	Hochschule RheinMain, Germany
Sahar Vahdati	University of Bonn, Germany
Anna Lisa Gentile	IBM Research Almaden, USA
Luigi Asprino	University of Bologna and STLab (ISTC-CNR), Italy
Valentina Presutti	STLab (ISTC-CNR), Italy
Mathieu d'Aquin	Knowledge Media Institute, The Open University
Muhammad Ismail	Jönköping University, Sweden
Luca Ferrari	University of Milan, Italy
Pasquale Lisena	Eurecom
Jinlong Guo	University of Illinois at Urban-Champaign, USA
Hala Skaf	Nantes University, France
Francesco Osborne	Knowledge Media Institute, The Open University
Aldo Gangemi	Université Paris 13 and CNR-ISTC, France
Maria Esther Vidal	Universidad Simon Bolivar, Venezuela
Marta Sabou	Vienna University of Technology, Austria
Axel Polleres	Vienna University of Economics and Business — WU Wien, Austria
Claudia D'Amato	University of Bari, Italy
Valentina Tamma	University of Liverpool, UK
Lora Aroyo	VU University Amsterdam, The Netherlands
Vanessa Lopez	IBM Research
Sai Gollapudi	International Institute of Information Technology, Telangana, India

Posters and Demos

Gregoire Burel	The Open University
Pierre-Antoine Champin	LIRIS
Mari Carmen Suárez-Figueroa	Universidad Politecnica de Madrid, Spain
Tudor Groza	Garvan Institute of Medical Research, Australia

Contents

Posters and Demos

Doctoral Consortium

Tutorial

Modeling, Generating, and Publishing Knowledge as Linked Data

Anastasia Dimou$^{(\boxtimes)}$, Pieter Heyvaert, Ruben Taelman, and Ruben Verborgh

Ghent University – imec – IDLab, Ghent, Belgium
{anastasia.dimou,pheyvaer.heyvaert,ruben.taelman,ruben.verborgh}@ugent.be

Abstract. The process of extracting, structuring, and organizing knowledge from one or multiple data sources and preparing it for the Semantic Web requires a dedicated class of systems. They enable processing large and originally heterogeneous data sources and capturing new knowledge. Offering existing data as Linked Data increases its shareability, extensibility, and reusability. However, using Linking Data as a means to represent knowledge can be easier said than done. In this tutorial, we elaborate on the importance of semantically annotating data and how existing technologies facilitate their mapping to Linked Data. We introduce [R2]RML languages to generate Linked Data derived from different heterogeneous data formats –e.g., DBs, XML, or JSON– and from different interfaces –e.g., files or Web APIs. Those who are not Semantic Web experts can annotate their data with the RMLEditor, whose user interface hides all underlying Semantic Web technologies to data owners. Last, we show how to easily publish Linked Data on the Web as Triple Pattern Fragments. As a result, participants, independently of their knowledge background, can model, annotate and publish data on their own.

Keywords: Linked Data generation · Linked Data publishing · [R2]RML · Linked Data Fragments

1 Introduction

Semantic Web technologies offer the possibility to integrate domain-level information from a combination of heterogeneous data sources and published as Linked Data. Thereby, Linked Data gains traction as a prominent solution for machine-interpretable knowledge representation. However, only a limited amount of data is available as Linked Data, as acquiring semantically enriched representations remains complicated, in addition to scalability issues that emerge once Linked Data is published for consumption.

This tutorial aims to show to data owners, who are domain-experts, how to perform the different steps to make their data available as Linked Data, namely modeling, generation, and publication. These steps altogether form a *Linked Data publishing workflow*. By the end of this tutorial, data owners should know how to profit of modeling the knowledge that appears in their data, semantically annotating them to generate corresponding Linked Data and publishing them.

© Springer International Publishing AG 2017
P. Ciancarini et al. (Eds.): EKAW 2016 Satellite Events, LNAI 10180, pp. 3–14, 2017.
DOI: 10.1007/978-3-319-58694-6_1

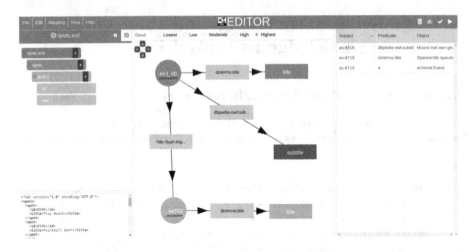

Fig. 1. The graphical user interface of the RMLEditor

The paper is organized as follows: in Sect. 2, we describe how mapping rules can be defined to specify the generation of Linked Data from raw data. In Sect. 3, we explain how such mapping rules can be executed to actually generate a Linked Data set. In Sect. 4 we discuss the publication of Linked Data. In Sect. 5, the administration of this workflow is described. Last, in Sect. 6 we explain how the tutorial was given at EKAW2016 conference.

2   Linked Data Modeling

Modeling is the first step of a Linked Data publishing workflow. It involves defining how to make knowledge available as Linked Data. In this step, raw data is modeled and semantically annotated using vocabularies. Data owners indicate:

- the **entities** that appear in the dataset, by assigning *IRIs* to them;
- how the **attributes** are related to the entities, by using *predicates*;
- what the **(data)types** are of these entities, by using *classes*, and of these attributes, by using **xsd**[1] or custom *datatypes*; and
- **relationships** between entities which might originally be in different data sources, by using *predicates*.

Mapping languages allow to declaratively specify the mapping rules which are defined during the modeling step. However, directly editing them using these languages is difficult for data owners who are not Semantic Web experts. Therefore, graphical user interface tools, such as the RMLEditor [14], are developed to ease modeling and support in defining mapping rules to semantically annotate raw data and generate Linked Data by hiding the underlying mapping language.

---

[1] https://www.w3.org/TR/swbp-xsch-datatypes/.

The RMLEditor [14] is a tool that enables data owners to define mapping rules that specify how Linked Data is generated using a uniform user interface that presents the data sources, mapping rules, and resulting Linked Data. It is implemented according to the desired features for mapping editors [15]. It is independent of the underlying mapping language, so data owners define mapping rules without knowledge of the language's syntax. Thus, even though the mapping rules execution might be performed directly in the tool for a sample of data, it allows to be exported and, hence, further processed, validated or executed.

The RMLEditor enables users to consider multiple data sources at the same time, as data might be spread across multiple sources. Moreover, it supports data in different formats, as the raw data might reside in, e.g., CSV, JSON, and XML files. As multiple vocabularies can be used, the RMLEditor supports the use of both existing and customs vocabularies. It enables multiple alternative modeling approaches [16], as certain use cases might benefit from using a specific approach and non-linear workflows, thus users are able to keep an overview of the different components which are involved.

The graphical user interface consists of three panels: Input, Modeling, and Results which are aligned next to each other from left to right as seen in Fig. 1. The **Input Panel** shows the data sources, both its structure and a sample of its raw data, and each data source is assigned with a unique color. The **Modeling Panel** presents the mapping rules using a graph-based visualization. The color of each node and edge depends on the data source that is used in that specific rule, if any. The panel offers the means to manipulate the nodes and edges of the graphs to update the rules. The **Results Panel** shows the resulting Linked Data when the mapping rules are executed on the data sources. For each RDF triple of the dataset, it shows the subject, predicate, and object.

The functionality of and the interaction between the panels supports different Linked Data generation approaches that depend on the use case and the data owners' preferences [15]. The *data-driven* approach uses the input data sources as the basis to construct the mapping rules. The classes, properties and datatypes of the schemas are then assigned to the mapping rules. The *schema-driven* approach occurs when data owners start with the vocabularies to define the mapping rules and data values from the data sources fill in the schema. Next, data fractions from the data sources can be associated with the mapping rules.

Semantic annotations can be applied relying on multiple vocabularies. The Linked Open Vocabularies[2] is integrated and can be consulted to get suggestions on which classes, properties and datatypes to use. As the graphs offer a generic representation of the mapping rules, they do not depend on the underlying mapping language. Additionally, the graph visualization and the mapping rules can be exported, allowing the execution of the mapping rules outside the RMLEditor. Additionally, by not restricting when users can interact with which panels, the RMLEditor supports *non-linear workflows*.

---

[2] http://lov.okfn.org/dataset/lov/.

3   Linked Data Generation

The next step in a typical Linked Data life cycle is the *generation* step. Although more and more data is semantically annotated and published as Linked Data, efficiently extracting and integrating information from diverse, distributed and heterogeneous sources to enable Semantic Web based applications remains complicated, let alone keeping track of their provenance. Moreover, the quality and consistency of the resulting Linked Data significantly varies, ranging from expensively curated to relatively low quality Linked Data sets.

Mapping languages enable detaching the mapping rules from the implementation that executes them and specifying in a declarative way how Linked Data is generated from raw data. User interfaces, as the aforementioned RMLEditor, allow defining mapping rules. However, dedicated tools *validate* and *execute* these to generate the desired Linked Data and assure their high quality.

3.1   Generation

Despite the significant number of existing tools, generating Linked Data from data in multiple heterogeneous sources, formats and access interfaces remains complicated. There is still no recommended formalization to define mapping rules regarding how to generate Linked Data derived from such raw data in an integrated and interoperable fashion, except for databases.

More precisely, the R2RML mapping language [4] was recommended by W3C in 2012 to define mapping rules for generating Linked Data but only from data which is derived from relational databases. In 2014, the RDF Mapping language RML [8] was proposed as a superset of the W3C-recommended mapping language R2RML, extending its applicability and broadening its scope.

RML is a generic mapping language defined to specify customized mapping rules that generate Linked Data from heterogeneous data in a uniform and integrated way. It is easily extended to cover references to different data structures, combined with case-specific extensions, but remains backward compatible with and follows the same syntax as R2RML. The RML vocabulary namespace is http://semweb.mmlab.be/ns/rml with a suggested prefix of **rml**. More details about RML can be found at http://rml.io/.

RML, in contrast to most mapping languages, also considers diverse dataset and services descriptions to access the raw data which is required to generate the desired Linked Data [9]. Corresponding vocabularies which are used to describe how to access the underlying raw data is aligned with the mapping rules definition. Such vocabularies may refer to: (i) the dataset's metadata (e.g., DCAT [3]); (ii) hypermedia-driven Web APIs (e.g., Hydra[4]); (iii) SPARQL services, (e.g., SPARQL-SD [5]); and (iv) database connectivity frameworks.

---

[3] https://www.w3.org/TR/vocab-dcat/.

[4] http://www.hydra-cg.com/spec/latest/core/.

[5] https://www.w3.org/TR/sparql11-service-description/.

Provenance and other metadata are essential to determine Linked Data set's ownership and trust. Thus, capturing them on every step of the Linked Data publishing workflow is systematically and incrementally required. Thus, declarative and machine-interpretable data descriptions, such as the aforementioned, and mapping rules, such as RML statements, are considered to automatically assert provenance and metadata information related to Linked Data generation [5].

3.2 Validation

Linked Data validation aims at aiding the data owners to acquire high quality Linked Data. The most frequent violations are related to the dataset's schema, namely the vocabularies used to annotate the original data [18]. The dataset's schema, on its own turn, derives from the classes and properties specified in mapping rules, while combining vocabularies increases the likelihood of violations to appear. Hence applying such mapping rules to raw data derived from one or more data sources, results in same violations being repeatedly observed even within the same Linked Data set.

Only recently efforts focus on formalizing Linked Data quality tracking and assessment [27]. Nevertheless, most approaches remain independent of the Linked Data workflow. In this tutorial, we follow a methodology [7] that incorporates systematically the Linked Data validation in the Linked Data workflow. A set of test cases ensures that the same violations are prevented to appear repeatedly within a Linked Data set and over distinct entities, and data publishers can discover violations even before the Linked Data is generated.

This methodology allows to uniformly validate both the mapping rules and the resulting Linked Data. To achieve that, we take advantage of mapping rules, such as RML statements, which are also expressed as Linked Data. Even though mapping rules specify how Linked Data will be formed and can cover many violations related to vocabularies which are used to annotate the data, some schema-related violations depend on how the mapping rules are instantiated on the original data. Thus, validating both the mapping rules and the resulting Linked Data ensures higher quality Linked Data.

We consider two tools in the frame of this tutorial that enable high quality Linked Data generation: (i) the **RMLProcessor** that executes mapping rules expressed in RML to generate Linked Data, and (ii) the **RMLValidator** that validates sets of mapping rules expressed in RML or Linked Data.

3.3 RMLProcessor

The RMLProcessor[6] is a tool that enables data owners to execute mapping rules which are defined as RML statements and generate the desired Linked Data. It requires a mapping document as input that summarizes the mapping rules to be executed. Those mapping rules contain also the references to the raw data which are considered to generate the desired Linked Data. The data owners may

---

[6] https://github.com/RMLio/RML-Mapper.

choose the preferred RDF serialization when the RMLProcessor is triggered and decide on executing all the mapping rules or a few of them.

Moreover, data owners can automatically generate the corresponding metadata and provenance. The desired metadata can be configured by a data owner who can define the vocabulary to be used and the desired details level. There are different details levels, the data owners may choose among: (i) dataset level, (ii) named graph level, (iii) partitioned dataset (iv) RDF triple level, or (v) RDF term level. The RMLProcessor considers implicit graphs for metadata which are related with the whole dataset, as well as for named graphs and reification triples if the metadata level is set on RDF triple or term level.

3.4 RMLValidator

The RMLValidator[7] [7] is a tool that enables data owners to validate both their mapping rules that are defined as RML statements and the Linked Data which are generated. It builds on RDFUnit [18] a validation framework for Linked Data. RDFUnit relies its function on pattern-based SPARQL templates that form the different test cases. Those test cases are instantiated for each vocabulary that is used to semantically annotate the data.

As [R2]RML mapping rules can be validated as Linked Data, because they are written as the generated triples and they are defined as Linked Data too. Thus, the same set of schema validation patterns normally applied on the Linked Data is also applicable on the mapping rules that state how the corresponding Linked Data is generated. The data owners need to provide to the RMLValidator (i) a mapping document with RML statements, or (ii) a Linked Data set and they receive a report with the validation results.

4   Linked Data Publishing

The next step in a typical Linked Data publishing workflow is *publication*. In this section, we discuss how Linked Data can be published. We presented the different Linked Data interfaces (Sect. 4.1); and non-technical tasks related to publishing (Sect. 4.2).

4.1   Linked Data Interfaces

One of the Linked Data publication main goals is for data to be *retrieved* and *discovered* by machines through HTTP interfaces. Not only the data, but also the interface(s) through which it is published, should be machine-understandable. *Linked Data Fragments* (LDF) [26] was introduced as a framework for comparing different Linked Data publication interfaces on dimensions such as the required server and client effort for querying them. It achieves this by conceptualizing the response of *any* HTTP interface for reading RDF as a *Linked Data Fragment*. Furthermore, this axis has been extended with a data dynamicity axis for

---

[7] https://github.com/RMLio/RML-Validator.

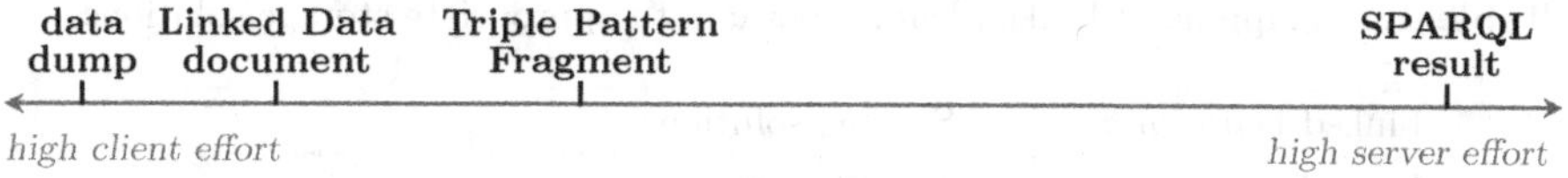

data dump	Linked Data document	Triple Pattern Fragment		SPARQL result

high client effort   *high server effort*

Fig. 2. Conceptual axis showing the server effort needed to query different types of Linked Data Fragments interfaces.

different types of temporal publication interfaces [21], such as TPF-QS [22] and CSPARQL [2]. Next, we discuss 4 types of interfaces using the LDF framework, as shown in Fig. 2.

Data Dump. A data dump is a file which contains a Linked Data set that is created once, and typically does not change anymore afterwards. These files can be available in any RDF serialization format, such as Turtle, RDF/XML, or TriG. Such data dumps may easily have sizes in the order of gigabytes. In these cases they may be available in a compressed archive such as gzip or HDT [11]. Data dumps do not provide any querying functionality themselves, it requires clients to download the data dump and load it locally if desired to query it. Hence, significant effort is required from the client to query such Linked Data sets, although little effort from the data owner is required.

Linked Data Document. The third Linked Data principle says "When someone looks up a URI, provide useful information, using the open Web standards such as RDF, SPARQL" [3]. When a URI is *dereferenced*, the provided information can be returned as a Linked Data document, containing information about the provided URI. This means that data becomes available in smaller fragments, which can be access on a per-subject basis. Such documents require a bit more effort from the publisher when compared to data dumps, because data needs to be available in fragments. Publication involves low server cost, browsing is easy, and querying such documents is possible by traversing links. However, query result completeness depends on the number of links within the data [13, 25].

SPARQL Query Result. SPARQL endpoints are a popular method for exposing Linked Data through an interface [10] that supports queries in the SPARQL query language [12]. Even though they are widely used across the Web, they suffer from significant availability issues [20]. The potentially high complexity of SPARQL queries, together with the public nature of query engines on the Web cause a very high load on servers that expose SPARQL access, which can lead to downtime. This makes SPARQL query engines a costly approach for publishing Linked Data. However, the required effort for clients to query these endpoints is very low, because the server performs the entire query evaluation process.

Triple Pattern Fragments. The Triple Pattern Fragments interface (TPF) [26] was introduced as a trade-off between server and client effort for querying. The approach consists of a low-cost server interface that accepts triple pattern queries, while clients then evaluate more complex SPARQL queries. Clients

Table 1. A mapping of Linked Data interfaces to appropriate storage solutions.

Linked Data interface	Storage solution
Data dump	RDF file, HDT, ...
Linked Data documents	Static or dynamically generated RDF files
TPF	RDF file, HDT, SPARQL engine, ...
SPARQL endpoint	SPARQL engine

achieve this by splitting up their queries into one or more triple pattern queries, sending them to a TPF interface, and joining their results locally. Experiments show that the TPF requires less effort from the server when compared to SPARQL endpoints [26], at the cost of slower query execution times and increase bandwidth. This approach makes it possible to publish Linked Data at a low cost while still enabling efficient querying, as illustrated by the high availability of DBpedia's TPF endpoint [24].

Depending on the desired interface, different storage solutions might be chosen. Table 1 shows what storage solutions can be used per Linked Data interface.

4.2   Linked Data Licensing, Announcement and Maintenance

In addition to the technical tasks of setting up an interface with an appropriate storage solution, there are also several non-technical tasks related to Linked Data publishing process, such as licensing and announcement which are performed following the best practises for publishing Linked Data [17].

Licensing. In most cases, the goal of Linked Data publishing is to reach Linked *Open* Data. By default, non-licensed data is not open, because regular copyright rules apply. The Open Knowledge Foundation[8] argues that openness is defined by (i) the availability and access of data; (ii) the possibility for anyone to reuse and redistribute the data; and (iii) universal participation, namely that no one should be excluded from these rights. A popular open license is the CC0 license[9], which is in line with the aforementioned definition of openness. To make the license known, it should be mentioned in the dataset listings and as metadata in the dataset using for example the Creative Commons vocabulary[10]. Some cases may require a non-open data license, such as datasets that include confidential data. The publication strategy should then support this license by applying security to the interface. This can be done by adding an authentication and authorization layer to the data-access interface for confidential and private information.

Announcement. After the infrastructure for publishing has been set up, the data is ready and properly licensed, its existence must be announced to the public. Depending on the target audience, different communication channels may

---

[8] https://okfn.org/opendata/.
[9] https://creativecommons.org/publicdomain/zero/1.0/.
[10] https://creativecommons.org/ns.

be of interest, which may include mailing lists, blogs, newsletters. During the announcement, a feedback channel must be in place. This allows data consumers to report issues or ask questions about the Linked Data set, when, for instance, the server would go down. Furthermore, the dataset can also be published on registries, such as https://datahub.io/. If the dataset has been properly annotated using dataset vocabularies such as DCAT [19] and VoID [1], crawlers may automatically discover and index a Linked Data set.

5   Linked Data Publishing Workflow Administration

A typical Linked Data workflow consists of the following steps: (i) *modeling* domain knowledge by defining mapping rules that specify how Linked Data may be generated; (ii) *generating* Linked Data by executing the mapping rules; and (iii) *publishing* Linked Data by turning it available for consumption.

Nevertheless, after Linked Data publication, the workflow is not over yet. In fact, it can be seen as a continuous process. Once a data publisher makes new data available, a *social contract* with data consumers is initiated. Data consumers depend on the availability of the published Linked Data, which needs to be guaranteed. This means that dataset and interface removal should be avoided at all costs. In practice, Linked Data sets will rarely remain static over time [23]. Detected violations or inconsistencies, as well as changes to the original raw data needs to be reflected in the Linked Data set.

Ideally, this is accomplished by publishing a new version without having to remove the old one. A good URI strategy should allow multiple dataset versions to exist next to each other. Linked Data set removal should be properly announced and the Linked Data set should then be moved to a different location. Redirects should be put into place, so as existing consumer applications will still work. Finally, an agent should be assigned as responsible for the Linked Data set maintenance and the feedback channel that also needs to be kept open.

We consider the **RMLWorkbench** in the frame of this tutorial that enables administrating and maintaining the complete Linked Data publication workflow, offering a user friendly interface to data owners.

5.1   RMLWorkbench

The **RMLWorkbench** [6] enables user friendly administration of the complete Linked Data publishing workflow in a single place. It offers a graphical user interface consisting of five panels: Access, Retrieve, Generate, Publish, and Schedule.

The **Access Panel** allows data owners to manage the different data sources which are considered to generate Linked Data. It supports data sources that are accessible through (i) local files, (ii) databases or (iii) the Web.

The **Retrieve Panel** enable data owners to specify which data from each data source from the Access Panel is considered for the Linked Data generation.

The **Generate Panel** gives data owners access to different mapping rules. The data owners have several methods to provide mapping rules: (i) upload a mapping document, (ii) specify a Web source with mapping rules, or (iii) create and edit rules via the RMLEditor. These rules can be executed, and the resulting Linked Dataset is available in the Publish Panel.

The **Publish Panel** panel enables data owners to publish the generated Linked Data via an LDF server. Nevertheless, the administrator can easily configure other interfaces, for instance a SPARQL endpoint. The data owners can then choose one or more of them to publish their Linked Data.

Last, the **Schedule Panel** enables users to specify the recurrence of the generation and publication steps. This allows to re-generate the desired Linked Data and keep it up-to-date with the original data.

6   EKAW2016 Tutorial Report

This tutorial was given at the 20th International Conference on Knowledge Engineering and Knowledge Management (EKAW2016). It lasted a whole day and consisted of three parts: (i) a broader introduction on Linked Data generation and publishing; (ii) a more specific introduction on [R2]RML for Linked Data generation and TPF for Linked Data publishing; and (iii) a practical session where participants used the RML tools to model and generate Linked Data and the TPF server and client to publish and query respectively the Linked Data which they generated. The former two were in the morning session, while the practical part was during the afternoon session. The detailed scheduled and the accompanying material are available at http://rml.io/EKAW2016tutorial.html. There were approximately twenty participants who are knowledge management or data experts. They were interested in using and profiting of Semantic Web technologies but still faced barriers to publish their own Linked Data.

Acknowledgements. The described research activities were funded by Ghent University, imec, the Flanders Innovation & Entrepreneurship (AIO), the Research Foundation – Flanders (FWO), and the European Union.

References

1. Alexander, K., Cyganiak, R., Hausenblas, M., Zhao, J.: Describing Linked Datasets with the VOID Vocabulary. Interest Group Note, W3C, March 2011. https://www.w3.org/TR/void/
2. Barbieri, D.F., Braga, D., Ceri, S., Valle, E.D., Grossniklaus, M.: Querying RDF streams with CSPARQL. Sigmod Rec. **39**(1), 20–26 (2010)
3. Berners-Lee, T.: Linked Data, July 2006. http://www.w3.org/DesignIssues/LinkedData.html
4. Das, S., Sundara, S., Cyganiak, R.: R2RML: RDB to RDF Mapping Language. W3C recommendation, W3C (2012). http://www.w3.org/TR/r2rml/
5. Dimou, A., De Nies, T., Verborgh, R., Mannens, E., Van de Walle, R.: Automated metadata generation for Linked Data generation and publishing workflows. In: Auer, S., Berners-Lee, T., Bizer, C., Heath, T. (eds.) Proceedings of the 9th Workshop on Linked Data on the Web. CEUR Workshop Proceedings, vol. 1593, April 2016

6. Dimou, A., Heyvaert, P., Maroy, W., De Graeve, L., Verborgh, R., Mannens, E.: Towards an interface for user-friendly Linked Data generation administration. In: Kawamura, T., Paulheim, H. (eds.) Proceedings of the 15th International Semantic Web Conference: Posters and Demos. CEUR Workshop Proceedings, vol. 1690, October 2016

7. Dimou, A., Kontokostas, D., Freudenberg, M., Verborgh, R., Lehmann, J., Mannens, E., Hellmann, S., Walle, R.: Assessing and refining mappingsto RDF to improve dataset quality. In: Arenas, M., et al. (eds.) ISWC 2015. LNCS, vol. 9367, pp. 133–149. Springer, Cham (2015). doi:10.1007/978-3-319-25010-6_8

8. Dimou, A., Vander Sande, M., Colpaert, P., Verborgh, R., Mannens, E., Van de Walle, R.: RML: a generic language for integrated RDF mappings of heterogeneous data. In: Proceedings of the 7th Workshop on Linked Data on the Web. CEUR Workshop Proceedings, vol. 1184, April 2014

9. Dimou, A., Verborgh, R., Vander Sande, M., Mannens, E., Van de Walle, R.: Machine-interpretable dataset and service descriptions for heterogeneous data access and retrieval. In: Proceedings of the 11th International Conference on Semantic Systems, pp. 145–152, September 2015

10. Feigenbaum, L., Todd Williams, G., Grant Clark, K., Torres, E.: SPARQL 1.1 Protocol. Recommendation (2013). http://www.w3.org/TR/sparql11-protocol/

11. Fernández, J.D., Martínez-Prieto, M.A., Gutiérrez, C., Polleres, A., Arias, M.: Binary RDF representation for publication and exchange (HDT). Web Semant. Sci. Serv. Agents World Wide Web **19**, 22–41 (2013)

12. Harris, S., Seaborne, A., Prud'hommeaux, E.: SPARQL 1.1 Query Language. Recommendation, W3C (2013). http://www.w3.org/TR/sparql11-query/

13. Hartig, O., Bizer, C., Freytag, J.-C.: Executing SPARQL queries over the web of Linked Data. In: Bernstein, A., Karger, D.R., Heath, T., Feigenbaum, L., Maynard, D., Motta, E., Thirunarayan, K. (eds.) ISWC 2009. LNCS, vol. 5823, pp. 293–309. Springer, Heidelberg (2009). doi:10.1007/978-3-642-04930-9_19

14. Heyvaert, P., Dimou, A., Herregodts, A.-L., Verborgh, R., Schuurman, D., Mannens, E., Walle, R.: RMLEditor: a graph-based mapping editor for Linked Data mappings. In: Sack, H., Blomqvist, E., d'Aquin, M., Ghidini, C., Ponzetto, S.P., Lange, C. (eds.) ESWC 2016. LNCS, vol. 9678, pp. 709–723. Springer, Cham (2016). doi:10.1007/978-3-319-34129-3_43

15. Heyvaert, P., Dimou, A., Verborgh, R., Mannens, E., Van de Walle, R.: Towards a uniform user interface for editing mapping definitions. In: Proceedings of the 4th Workshop on Intelligent Exploration of Semantic Data. CEUR Workshop Proceedings, vol. 1472, October 2015

16. Heyvaert, P., Dimou, A., Verborgh, R., Mannens, E., Van de Walle, R.: Towards approaches for generating RDF mapping definitions. In: Proceedings of the 14th International Semantic Web Conference: Posters and Demos. CEUR Workshop Proceedings, vol. 1486, October 2015

17. Hyland, B., Atemezing, G., Villazón-Terrazas, B.: Best practices for publishing Linked Data. Working Group Note, W3C, January 2014. http://www.w3.org/TR/ld-bp/

18. Kontokostas, D., Westphal, P., Auer, S., Hellmann, S., Lehmann, J., Cornelissen, R., Zaveri, A.: Test-driven evaluation of Linked Data quality. In: Proceedings of the 23rd International Conference on World Wide Web (2014)

19. Maali, F., Erickson, J.: Data Catalog Vocabulary (DCAT). W3C Recommendation, W3C (2014). http://www.w3.org/TR/vocab-dcat/

20. Pérez, J., Arenas, M., Gutierrez, C.: Semantics and complexity of SPARQL. In: Cruz, I., et al. (eds.) ISWC 2006. LNCS, vol. 4273, pp. 30–43. Springer, Heidelberg (2006). doi:10.1007/11926078_3

21. Taelman, R.: Continuously self-updating query results over dynamic heterogeneous Linked Data. In: Sack, H., Blomqvist, E., d'Aquin, M., Ghidini, C., Ponzetto, S.P., Lange, C. (eds.) ESWC 2016. LNCS, vol. 9678, pp. 863–872. Springer, Cham (2016). doi:10.1007/978-3-319-34129-3_55

22. Taelman, R., Verborgh, R., Colpaert, P., Mannens, E.: Continuous client-side query evaluation over dynamic Linked Data. In: Sack, H., Rizzo, G., Steinmetz, N., Mladenić, D., Auer, S., Lange, C. (eds.) ESWC 2016. LNCS, vol. 9989, pp. 273–289. Springer, Cham (2016). doi:10.1007/978-3-319-47602-5_44

23. Umbrich, J., Decker, S., Hausenblas, M., Polleres, A., Hogan, A.: Towards dataset dynamics: change frequency of Linked Open data sources. In: Proceedings of the WWW 2010 Workshop on Linked Data on the Web. CEUR Workshop Proceedings, vol. 628, April 2010

24. Verborgh, R.: DBpedia's triple pattern fragments: usage patterns and insights. In: Gandon, F., Guéret, C., Villata, S., Breslin, J., Faron-Zucker, C., Zimmermann, A. (eds.) ESWC 2015. LNCS, vol. 9341, pp. 431–442. Springer, Cham (2015). doi:10. 1007/978-3-319-25639-9_54

25. Verborgh, R.: Piecing the puzzle - self-publishing queryable research data on the Web. In: Proceedings of the 10th Workshop on Linked Data on the Web (2017)

26. Verborgh, R., Vander Sande, M., Hartig, O., Van Herwegen, J., De Vocht, L., De Meester, B., Haesendonck, G., Colpaert, P.: Triple pattern fragments: a low-cost knowledge graph interface for the web. J. Web Semant. **37–38**, 184–206 (2016)

27. Zaveri, A., Rula, A., Maurino, A., Pietrobon, R., Lehmann, J., Auer, S.: Quality assessment for linked data: a survey. Semant. Web J. (2015)

First Workshop on Detection, Representation and Management of Concept Drift in Linked Open Data: Report of the Drift-a-LOD2016 Workshop

Detection, Representation and Management of Concept Drift in Linked Open Data: Report of the Drift-a-LOD2016 Workshop

Laura Hollink[1(✉)], Sándor Darányi[2], Albert Meroño-Peñuela[3],
and Efstratios Kontopoulos[4]

[1] Centrum Wiskunde & Informatica, Amsterdam, The Netherlands
l.hollink@cwi.nl
[2] University of Borås, Borås, Sweden
[3] Vrije Universiteit Amsterdam, Amsterdam, The Netherlands
[4] Centre for Research & Technology Hellas, Thessaloniki, Greece

1 Introduction

The Web of Data has expanded to – and is being deployed in – a wide range of domains. In many of these domains, facts change continuously: political landscapes evolve, medical discoveries lead to new cures, and artists form new collaborations. In terms of knowledge representation, we observe that instances change their roles, new relations appear, old ones become invalid, and classes change both their definition and member-instances. These changes pose new challenges to creators and users of Linked Open Data (LOD) to avoid that semantic interoperability and access to digital content are compromised. For instance, LOD publishers need to detect changes in the real world and capture them in their datasets; users and applications need automated tools to adapt querying over such diachronic datasets.

Whereas the Semantic Web field has been concerned largely with static knowledge for a long time, there is now more and more interest in time-dependent content. Event modeling, event extraction, stream reasoning, and the emergence of Web Observatories are examples of that. While the infrastructure to represent and query time-stamped data is coming together, there is still a lack of knowledge about how to detect that facts and concepts have changed, how to interpret changes, and how to deliver results to users in a meaningful way. Throughout Europe, researchers are currently working on these issues. The DIACHRON project[1] addresses the problem of versioning and evolution of LOD; PERICLES[2] focuses on the disclosure of digital content in changing environments. In the ELISA project[3], the impact of evolving ontologies on semantic annotations is investigated, while the recently finalized PRELIDA[4] project aimed to connect the LOD community to the Digital Preservation community, to promote work on the

---

[1] http://www.diachron-fp7.eu/.
[2] http://pericles-project.eu/.
[3] https://www.list.lu/en/project/elisa/.
[4] http://www.prelida.eu/.

preservation of LOD and to guarantee long-term access. These efforts show that there is a growing interest in this topic not only from the Semantic Web community but also from the Databases, Digital Preservation and Digital Libraries communities.

2 Workshop Overview

Drift-a-LOD2016, the first workshop about concept drift and Linked Open Data, was held at the 20th International Conference on Knowledge Engineering and Knowledge Management (EKAW) in Bologna, Italy, on November 20, 2016. The workshop brought together a diverse range of researchers working on detecting, representing and managing the fact that the world changes. Being from different communities, participants used different terminology to refer to this phenomenon (or related phenomena), speaking about concept drift, semantic change, meaning change, lexical change, vocabulary shift, etc.

In particular, we distinguish contributions with a Knowledge Engineering focus, studying changes of concepts in structured knowledge, and contributions from an NLP perspective, studying language change in large corpora of natural language text. The synergy between the two communities was evident from various concrete examples. For instance, time-stamped knowledge graphs may help in the correct interpretation of natural language texts (e.g. word sense disambiguation in historic texts). The other way around, word change as observed through NLP methods in large text corpora can be used as a signal that a concept in a knowledge base has changed and needs to be updated.

The morning was devoted to 'NLP-inspired' work. In this session, Lea Frermann gave the first keynote of the day about the SCAN model of diachronic meaning change, quantifying and visualizing how the relative frequency of word-senses changes over time. SCAN is knowledge lean (i.e. does not depend on pre-processing tools such as parsers or entity detectors) and models change as a smooth rather than abrupt process [4].

In the afternoon, Cédric Pruski kicked off the session focused on Knowledge Engineering with a keynote about the ELISA project (mentioned above) and it's predecessor DYMAMO. He presented an empirical study of what types of change happen in a biomedical ontology, and how they impact existing semantic annotations [7]. Next to the two keynotes, the workshop included six regular presentations, two late breaking result pitches, and a discussion session in which directions for future research were proposed.

3 Research Directions

There is a lack of consensus about the various ways in which concepts change. At least two attempts have been made so far to make sense of it. Wang et al. [8] define three types of concept drift. They distinguish change of the label, the intension and the extension of a concept. During the workshop, Fokkens et al. [3] opposed the idea that extensional change is a form of concept drift. They proposed alternative definitions, noting that not all cases of intensional change are to be considered concept drift; if the core of the concept has changed, it is more suitable to think of it as a new concept than as a drifted concept.

Another open question is related to the mapping between lexical change and concept change. In other words, if a meaning of a word changes, what does that say about the concept(s) denoted by that word? Van Aggelen et al. [1] take a first step by connecting word-level change scores derived from word2vec models to synsets in WordNet.

Finally, the need for large-scale evaluation data was felt by all participants. The discussion highlighted several options, each with pros and cons. Firstly, small sets of human-judged change scores have been described (e.g. 21 concepts in [5]). The DBpedia wayback machine Fernández et al. [2] provides a wealth of data, but not a ground truth to distinguish which edits are to be considered concept drift, and which are simply corrections or additions to knowledge that was already there. As an alternative, Frermann and Lapata [4] use the SemEval task for dating text. Finally, Recchia et al. [6], in their workshop contribution, proposed a new method based on changes in the subject classifications of the British National Bibliography, which are published as Linked Open Data.

References

1. van Aggelen, A., Hollink, L., van Ossenbruggen, J.: Combining distributional semantics and structured data to study lexical change. In: Proceedings of the 1st Workshop on Detection, Representation and Management of Concept Drift in Linked Open Data, Bologna, Italy (2016)
2. Fernández, J.D., Schneider, P., Umbrich, J.: The DBpedia wayback machine. In: Proceedings of the 11th International Conference on Semantic Systems (SEMANTiCS 2015), Vienna, Austria, pp. 192–195. ACM (2015)
3. Fokkens, A., Braake, S.T., Maks, I., Ceolin, D.: On the semantics of concept drift: towards formal definitions of concept drift and semantic change. In: Proceedings of the 1st Workshop on Detection, Representation and Management of Concept Drift in Linked Open Data, Bologna, Italy (2016)
4. Frermann, L., Lapata, M.: A Bayesian model of diachronic meaning change. Trans. ACL **4**, 31–45 (2016)
5. Kenter, T., Wevers, M., Huijnen, P., de Rijke, M.: Ad hoc monitoring of vocabulary shifts over time. In: Proceedings of the 24th International Conference on Information and Knowledge Management, pp. 1191–1200 (2015)
6. Recchia, G., Jones, E., Nulty, P., Regan, J., de Bolla, P.: Tracing shifting conceptual vocabularies through time. In: Proceedings of the 1st Workshop on Detection, Representation and Management of Concept Drift in Linked Open Data, Bologna, Italy (2016)
7. dos Reis, J.C., Dinh, D., Silveira, M.D., Pruski, C., Reynaud-Delaître, C.: Recognizing lexical and semantic change patterns in evolving life science ontologies to inform mapping adaptation. Artif. Intell. Med. **63**(3), 153–170 (2015)
8. Wang, S., Schlobach, S., Klein, M.C.A.: Concept drift and how to identify it. J. Web Semant. **9**(3), 247–265 (2011)

Tracing Shifting Conceptual Vocabularies Through Time

Gabriel Recchia[✉], Ewan Jones, Paul Nulty, John Regan, and Peter de Bolla

The Concept Lab, CRASSH, University of Cambridge, Cambridge, UK
{glr29,ejj25,pgn26,jjr35,pld20}@cam.ac.uk

Abstract. This paper presents work in progress on an algorithm to track and identify changes in the vocabulary used to describe particular concepts over time, with emphasis on treating concepts as distinct from changes in word meaning. We apply the algorithm to word vectors generated from Google Books n-grams from 1800–1990 and evaluate the induced networks with respect to their *flexibility* (robustness to changes in vocabulary) and *stability* (they should not leap from topic to topic). We also describe work in progress using the British National Biography Linked Open Data Serials to construct a "ground truth" evaluation dataset for algorithms which aim to detect shifts in the vocabulary used to describe concepts. Finally, we discuss limitations of the proposed method, ways in which the method could be improved in the future, and other considerations.

Keywords: Concepts · Word embeddings · Linked open data

1 Introduction

Some influential theories of conceptual structure, such as the so-called *name priority view* [1] and some interpretations of the *classical theory of concepts* [2], treat concepts[1] as essentially in one-to-one correspondence to word senses [3–5]. On this view, one word might have several different senses and thereby correspond to several different concepts, but it is nonetheless possible to identify concepts via a careful examination of word meanings. Some modern philosophers and psychologists have made convincing arguments that this view is overly simplistic or flat-out wrong [1, 6]. Even if one does believe in a direct correspondence between word senses and concepts, however, it is clear that a change in word sense does not necessarily entail a change in the concept that was originally associated with it. For example, the word *broadcast*

---

[1] Rather than thinking of concepts in a way that strongly links them to a particular lexeme (e.g., "the concept of justice"), we have argued elsewhere that it is preferable to think of concepts (at least insofar as they are expressed in discourse) in terms of their functions, one of which is to permit two interlocutors to sense that they have arrived at a common understanding of the matter under discussion. This is rather different and more abstract than the notion of a concept as being equivalent to a class in a classical ontology, and more specific than a theme or topic. However, for purposes of clarity and compatibility with the way related work speaks about "concepts," our use of the word in this paper roughly conforms to the vague OED definition of "a general idea or notion." We are explicitly *not* using it to refer to "the meaning that is realized by a word or expression.".

© Springer International Publishing AG 2017
P. Ciancarini et al. (Eds.): EKAW 2016 Satellite Events, LNAI 10180, pp. 19–28, 2017.
DOI: 10.1007/978-3-319-58694-6_2

started to change from having the meaning of "scattering [seed] abroad over the whole surface, instead of being sown in drills or rows" to being associated with the transmission of radio or television signals in the 1920s [7, 8]. However, the fact that the primary sense of *broadcast* changed did not mean that the concept of sowing seeds over a wide area went away. Similarly, it seems clear that a culture could possess a particular concept even if no corresponding word or collocation exists in the primary language spoken by members of that culture.

The distinction between word senses and concepts is an important one to draw because, as pointed out by Wevers et al. [9], some computational approaches described as methods for detecting changes in "concepts" are often actually methods for detecting changes in the use of a single word or an unchanging group of words over time. Because word senses change over time, a change in the frequency or lexical associations of a particular word does not necessarily entail a change in the concept of interest. Being able to track concepts over time in a way that is robust to shifting vocabularies is therefore essential. Methods for detecting conceptual change in time-varying textual sources are particularly relevant in the context of Linked Open Data (LOD). To assist in the maintenance of LOD ontologies, knowledge engineers may wish to use time-varying text corpora, such as academic journals or news sources, to monitor conceptual change over time. Consider someone who maintains an ontology intended to represent relationships between various concepts in the neuroscience literature, who notices that this year there has been a marked uptick in the frequency of particular words that previously occurred only rarely. Does this merit the addition of a new class to the ontology? Or is this simply novel language for describing an old idea? Ultimately, this must come down to human judgment, but automatic methods for assisting with the decision could highlight important related classes already represented in the knowledgebase.

This paper presents work in progress toward an algorithm to track vocabulary associated with particular concepts over time in a flexible and stable way. In the next section, we describe related work, particularly a promising model recently developed by [10]. In Sect. 3 we implement a method which avoids one of the weaknesses of previous work while retaining the most important benefits. In Sect. 4, we describe work in progress using the *British National Biography Linked Open Data Serials* to construct a "ground truth" evaluation dataset for algorithms of this sort. We also discuss an additional approach to constructing a suitable ground truth dataset that we believe may hold some promise. Finally, we discuss some limitations of the method proposed in Sect. 3 and discuss other considerations. In particular, we highlight the fact that the present method has difficulty distinguishing between periods of "some vocabulary change" and "massive vocabulary change," and suggest a way to improve its capacity to do so. We also highlight limitations of focusing only on densely associated clusters of words when attempting to understand conceptual change quantitatively.

2   Related Work

To address the problem of word sense change described in the Introduction, the *Concepts Through Time* model advocates an alternative approach to tracking concepts, using a set of Dutch newspapers from 1890–1990 as a corpus [9]. Rather than selecting a static set of terms and monitoring its frequency over the entire century, they select an initial term or terms of interest and find a cluster of words that are highly similar, according to a word embedding model trained on a specific timeslice (e.g., articles from the years 1890–1900). The cluster is updated from timeslice to subsequent timeslice in a manner which acknowledges that "the set of words used to discuss a particular concept might not show any overlap at all between different periods of time" [9]. However, treating time-shifted collections of words with no overlap whatsoever as the 'same' has its own drawbacks. As [11] points out, "Imagine a subset of documents containing strong co-occurrence patterns across time: first between birds and aerodynamics, then aerodynamics and heat, then heat and quantum mechanics—this could lead to a single topic that follows this trajectory, and lead the user to inappropriately conclude that birds and quantum mechanics are time-shifted versions of the same topic." Perhaps for this reason, subsequent work by the developers of *Concepts Through Time* notes that "a successful system... should strike a balance between an adaptive strategy that responds to changes in vocabulary, and a more conservative approach that keeps the vocabulary stable" [10]. Their revised model requires a user to select an initial set of seed terms and an algorithm to construct vocabularies of related terms for each timeslice: *adaptive, nonadaptive,* or *hybrid.* A simplified description of their adaptive method is that it starts with an input vocabulary (initially the user's set of seed terms), expands that by adding words exceeding some minimum similarity threshold to the set, constructs a network from this set such that all pairs of nodes (words) exceeding the threshold are assigned an edge, and then prunes nodes that are low in degree centrality. The resulting word set is used as the input vocabulary for the next timeslice. This method works, but sometimes results in unacceptably high levels of drift away from the original concept. To counter this, they implement a nonadaptive method which simply outputs the words most similar to the words in the original seed set for every sliding time window, as well as a hybrid method which replaces "the least central terms of the vocabularies produced by the nonadaptive method by the top *i* vocabulary terms produced by the adaptive method with respect to degree centrality" [10]. This hybrid method performs best when compared against human judgments of performance.

Topic models have been another popular approach for monitoring groups of related words over time [11–14]. However, these often either do not explicitly model changes in vocabulary within a particular topic/concept, or do not pay explicit attention whether the method allows topics to drift far afield from their original conceptual content. One contribution of the present work is that it does both, while resolving an important difficulty with the most similar approach we are aware of.

Finally, much work has been done on automatically tracing changes in a given word's meaning over time, e.g. [8, 15, 16]. Although this clearly differs from our aim of tracing changes in the vocabulary used to describe particular concepts, these methods are extremely useful for our purposes. For example, a word may need to be excluded from a core of tightly interrelated terms if its meaning drifts too far afield from the rest. We therefore here make extensive use of the *HistWords* vectors [8] developed by applying skip-grams with negative sampling (one of the algorithms available in *word2vec*) to n-grams distributed by Google Books. *HistWords* contains a separate vector for each of a very large number of terms for every decade from 1800 to 1990, such that words that appear in similar contexts within a given timeslice have similar vectors. Such vectors successfully capture shifts in word meaning over time [8], and we use the same approach and data to quantify semantic similarity. We describe how we use these vectors in more detail in the following section.

3   Time-Varying Relationships in Text

Recall that the adaptive method of [10] involves an expansion step in which words related to any word in the input vocabulary are added to the network as nodes, and a pruning step in which nodes low in network centrality (in-degree or out-degree) are pruned. Although this is an excellent way to pull in novel vocabulary while also preventing the overall network from drifting too far afield, it has one unintended consequence. When the input vocabulary contains a word linked to two densely connected but unrelated clusters (e.g., a polysemous word), unrelated clusters of words will be added during the expansion step (Fig. 1). Because nodes in each cluster have high degree, they will not be eliminated in the pruning step. The consequence is that unrelated, weakly connected clusters can persist as part of the same "concept." The example in Fig. 1 makes this particularly clear by illustrating two clusters so unrelated that they would become disconnected if the node connecting them were pruned. However, it is important to recognize that this phenomenon remains a problem even if a constraint were imposed requiring the graph to be fully connected. The other two methods described by [10] (*nonadaptive* and *hybrid*) suffer from the same difficulty.

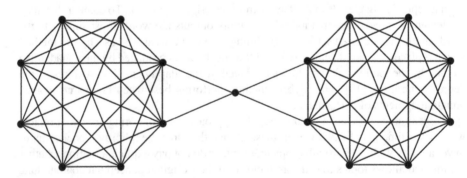

Fig. 1. A graph representing two unrelated clusters of words connected by a single polysemous word. Even if the center node is eliminated in the pruning step described in [10], the nodes in each 8-clique may not be, due to their high degree. Both clusters thus erroneously continue to be interpreted as part of the same "concept."

Our method addresses this by allowing two nodes to be treated as part of the same "conceptual network" only if all words in the network are highly related to all other words in the network. This constraint results in the happy consequence that words are excluded which may be highly related to one or two words in a particular network by happenstance, but are not close enough to the "conceptual core" to be related to each of the others. If a graph is constructed in which highly related words (nodes) are connected by edges, then such a "conceptual network" corresponds simply to the graph-theoretical notion of a clique. Because all subsequent descriptions of networks in this paper concern weighted graphs in which nodes represent words, edges represent strong corpus-based associations between words, and edge weight corresponds to the strength of the asso- ciation, we use "nodes" and "words" interchangeably, and "relatedness," "similarity," and "edge weight" as synonymous with "cosine similarity" (e.g., similarity between word vectors in the *HistWords* data). Like [10], we treat documents from every timeslice (in our case, decades from 1800–1990) as a separate subcorpus and build a separate vector space corresponding to each.

4 Algorithm

Given a size k and a seed set of words W, the algorithm begins by finding the fully connected subgraph of size k containing all words in W such that the minimum edge weight (in the earliest timeslice) is as high as possible (i.e., higher than in all other such subgraphs that meet the criteria). This can be accomplished efficiently by attempting to find a subgraph of size k containing all words in W such that every edge exceeds a very high threshold[2], but then gradually lowering the threshold until an appropriate subgraph is found. Afterwards, the vectors for the second timeslice are loaded, and the subgraph is updated by attempting to answer the question, "Is it possible to increase the minimum edge weight by replacing one of these nodes with a node currently not in the subgraph? If so, which of all possible replacements would increase the minimum edge weight the most?[3]" Because typically only one edge is equal to the current minimum edge weight, this can also be computed efficiently. This corresponds to a "drop one, add one" rule where, for any given timeslice, a single word from the network of the previous timeslice will be replaced if and only if doing so increases the minimum similarity between every word pair in the resulting network. The process repeats for every subsequent timeslice. Table 1 illustrates an example of an evolving network built using this method.

Our primary concerns were that conceptual networks be traced in such a way that is *flexible* (words whose meanings shift away from the conceptual core should drop out) but also *stable* (a network initially about birds should not drift to quantum mechanics). We tested the model by initializing it with 500 words randomly selected from the 30,000 most frequent terms in *HistWords*, which were used as the lexicon. Of these, there were 212 words such that a fully connected network of size 9 existed with a minimum edge weight of 0.2 or greater could be constructed.

A potential criticism of this model is that while it purports to be 'flexible' in the sense that it traces a group of *conceptually related* words (rather than merely words associated with the seed term alone), the fact that it is initialized with words closely related to the seed term may mean that in practice the seed term always ends up as a permanent part

---

[2] Because the threshold is initially set so high that no such subgraph can be found, this method ensures that the first subgraph discovered which meets these criteria is the one desired.

[3] Note that every node in the subgraph must correspond to a unique word.

of the network. Table 1 illustrates that in at least one case of radical semantic change (the word "gay"), the seed term does successfully drop out by the 1940s. However, it is possible that this virtually never happens. Flexibility was therefore evaluated by quantifying the proportion of the 212 initial networks (year 1800) in which the seed term did drop out by 1990. Another potential criticism is the reverse: Because every timeslice offers an opportunity to jettison one term and incorporate a new one, networks might drift to completely different topics. For example, if each iteration caused a random word to be replaced with a word not previously in the network, then after 19 timesteps a typical graph of size 9 would be expected to retain only $(8/9)^{19} = 10.7\%$ of its initial vocabulary. We therefore also computed the proportion of the vocabulary shared in the 1800 vs. 1990 clusters, with qualitative analysis of the clusters with the least shared vocabulary, to evaluate whether they exhibited less drift than this random baseline.

Table 1. Evolution over time of the network constructed from the seed word "gay."

1900	affable, cheerful, courteous, gay, genial, humored, natured, sprightly, witty
1910	affable, cheerful, courteous, gay, genial, humored, natured, humoured, witty
1920	affable, cheerful, courteous, gay, genial, jovial, natured, humoured, witty
1930	affable, cheerful, courteous, gay, mannered, jovial, natured, humoured, witty
1940	affable, cheerful, courteous, amiable, mannered, jovial, natured, humoured, witty
1950	affable, cheerful, courteous, amiable, mannered, vivacious, natured, humoured, witty
1960	affable, cheerful, courteous, amiable, mannered, charming, natured, humoured, witty
1970	affable, cheerful, courteous, amiable, mannered, charming, natured, gentle, witty
1980	affable, cheerful, courteous, amiable, humored, charming, natured, gentle, witty
1990	affable, cheerful, courteous, amiable, clever, charming, natured, gentle, witty

5 Results

With respect to flexibility, the seed word used to generate the initial size-9 network in 1800 was no longer present in the 1990 network in 147 of 212 cases (69%). In 91% of these cases, the seed word never re-entered the network once it had dropped out, suggesting that the seed word can indeed be permanently ejected from the conceptual core if its meaning or associations drift in a different direction. With respect to stability, the average overlap in vocabulary between the initial 1800s network and the final 1990 s network was 33%, with all 212 cases sharing at least one word (11%) in common with the original 9-word network. Even when only one word was shared, the network typically did not drift too far afield, as in the case of the seed word "uneasy" (1800: *anxieties, dejected, dejection, distraction, fits, insupportable, languishing, uneasy, weariness*; 1990: *anxieties, grief, despair, disappointment, misery, sorrow, anguish, sadness, loneliness*). The full set of networks generated in this evaluation may be obtained from http://nowin2d.com/vocabularies.html.

6 Discussion

The results suggest that even with such a rigid algorithm, the induced conceptual vocabularies are certainly flexible and reasonably resistant to drift. However, there were occasional clear cases in which vocabulary drifted significantly away from the original conceptual core, as in the case of the network generated from the seed word "logical" (1800: *abstruse, definitions, disquisition, disquisitions, explanations, explication, grammatical, illustrating, logical*), which drifted towards different areas of academic study by the 1990s (*abstruse, mathematical, philosophy, theory, metaphysics, metaphysical, empirical, theoretical, philosophical*). One obvious direction for future work is the optimization of initialization parameters (network size, initial edge weight threshold), which were chosen arbitrarily. However, the fact that even arbitrarily chosen initialization parameters resulted in reasonably flexible and stable networks is promising. Other directions could include making use of the information about how a conceptual vocabulary has changed in the past to predict how it will change in the future, using techniques that parallel those applied in predicting ontology evolution, e.g. [17]. In addition, future work should consider a stronger evaluation metric. The final section describes work in progress towards constructing a "ground truth" evaluation dataset that could be used to do just that.

7 Constructing Ground Truth Evaluation Data from LOD

To more fully evaluate an algorithm's ability to track vocabulary change associated with arbitrary concepts, a "ground truth" dataset is necessary. The only such data of which we are aware is the *Ground Truth Set for Monitoring Shifts in Vocabulary Over Time*[4] offered by the developers of Concepts Through Time [10]. However, this dataset is limited to 21 concepts spanning only four decades and is in Dutch, making it incompatible with most large, diachronic corpora.

However, something very near to a much larger, English-language ground truth dataset already exists as LOD, in the form of the British National Bibliography Linked Open Data (BNBLOD) collections. Although we see 'concepts' as nonidentical to subjects as defined by the BNB, it is nonetheless likely that there is a high level of conceptual relatedness between all documents to which the British Library has assigned the subject http://bnb.data.bl.uk/id/concept/lcsh/Engineering, even if the vocabulary of such documents differs markedly from year to year. Particularly useful are the BNBLOD Serials, which in addition to including the year of each journal's first publication, very commonly contain "Journal of X" in the title, where "X" corresponds to a short phrase describing a particular subject. The vocabulary of such 'title phrases' is often tied to a particular moment in time. Consider, for example, the phrases so extracted from the earliest "journal of X" journals in the BNBLOD Serials assigned the subject of *Psychiatry* (1876: 'nervous and mental disease'), *Engineering* (1921: 'applied mathematics and mechanics'), *Entrepreneurship* (1985: 'business venturing'), and *Tourism*

---

[4] Available: http://ilps.science.uva.nl/resources/shifts/.

(1972: 'travel research'). These phrases are no longer commonly used to describe these subjects. As a first step in constructing an evaluation dataset, therefore, we are first simply extracting title phrases, publication dates, and subjects from the BNBLOD Serials and structuring them as follows: Given a start year y_1, an end year y_2, and a phrase extracted from the title of a serial having subject S first published in y_1, the algorithm being evaluated must predict which words and phrases are most likely to appear in titles of other journals of subject S which were published in y_2. A robust algorithm trained on an appropriate could ideally correctly identify that the cluster of words that contains, e.g., "business venturing" in 1985 ought to include "entrepreneurship" by 1995 (rather than, say, "business organization"), and that "travel research" in 1972 is closer to "tourism" in 1992 than to "educational travel".

It should be noted that this is just a first step, and we hope that other, more comprehensive methods of evaluation will be developed with time. Possibilities that may inspire readers could include working with historians in the manner of [10] to create a larger dataset, or even using Library of Congress subject headings. Given that the same set of headings has been used to classify current works as well as now-digitized works going back as far as the 16th century (e.g., many of the volumes found in *Early English Books Online*), these subject headings could provide an interesting way to quantify changes in vocabulary over very large timespans. For example, of the nine works in *Early English Books Online* prior to 1700 which are tagged with subject headings containing "Biology", none contain the word "biology" in the full text [18]. This is not surprising, as the Oxford English Dictionary reports the first known use of the word as used in the scientific sense as dating from 1799 [19]. Standard methods from computational linguistics, such as *tf-idf*, could be employed to extract terms that are particularly prominent in texts published in particular decades labeled with particular Library of Congress subject keywords. This approach would essentially use the Library of Congress subject headings as a supervised 'ground truth' of what documents are related to a given concept (e.g. biology) from each timeslice, and by extension, the prominent words in these documents would be treated as the 'ground truth' with respect to the vocabulary that is associated with that concept at different times. It is our hope that the development of robust evaluation datasets will allow us not only to better evaluate our own research but move the field of representing diachronic conceptual change forward as a whole.

8   Limitations, Potential Improvements, Other Considerations

Although we were able to demonstrate that the proposed method is flexible (allows words that shift meaning to drop out of a lexical cluster) and resistant to semantic drift, the lack of a suitable ground truth dataset on which to objectively evaluate the degree to which the method tracks the vocabulary of a single "concept" across time is the most crucial limitation, as discussed in the previous section. Another important limitation stems from the decision to impose a rigid filter on the amount to which (the "drop one, add one" rule). The present method does not force a word to 'drop out,' so when there is little change in the degree to which a group of words cluster together from one decade to the next, the network will not change. However, because at most one word may drop out when change is present, it is impossible to distinguish between periods of gradual vs. rapid conceptual change using this approach. Allowing the addition of multiple words to a network in certain cases – for example, when all words in a particular clique became more highly associated with some set of n words than they are with each other, for some $n > 1$ – is an obvious way to remedy this. Of course, allowing more than one word to be added to a clique at once increases the chances of permitting unacceptable levels of drift, returning us to the central problem: evaluating whether or not a group of words was drifting overmuch from its corresponding concept would require a reliable ground truth dataset.

In addition, there are unanswered theoretical questions about whether tracking a tight cluster of associated words is tantamount to tracking a 'concept.' We suspect that it is not. If one's goal is to detect and understand conceptual change across time, it may well turn out that the objects of interest are something more akin to 'conceptual architectures': larger patterns that may include, but are not limited to, dense clusters of tightly bound lexis. For example, clusters may split and merge; two clusters of words that were previously unassociated may become indirectly associated by virtue of a new, mediating cluster that is linked to both; and some entities that are not clusters may hold important clues to conceptual change (e.g., single words that undergo significant drift, not because of changes in their meaning, but because of changes in their connotations, contexts of use, and so on). All of these considerations underscore the complexity and promise of novel approaches to tracking the vocabulary associated with particular concepts over time.

Acknowledgments. This paper is a revision of a paper that previously appeared in the 2016 CEUR Workshop Proceedings [20], which was invited to be included, after expansion and revision, into the present volume. The research presented here was supported by a private donation to the Cambridge Centre for Digital Knowledge (CCDK) at the University of Cambridge.

References

1. Seiler, T.B., Wannenmacher, W. (eds.): Concept Development and the Development of Word Meaning, vol. 12. Springer Science & Business Media, Berlin (2012)
2. Margolis, E., Laurence, S.: Concepts. In: Zalta, E.N. (ed.) The Stanford Encyclopedia of Philosophy (2014). http://plato.stanford.edu/archives/spr2014/entries/concepts/
3. Fodor, J.A.: The Language of Thought. Crowell, New York (1975)
4. Clark, E.V.: Meaning and concepts. In: Mussen, P.H. (ed.) Handbook of Child Psychology, vol. 3. Cognitive Development, pp. 787–840. Wiley, New York (1983)
5. Murphy, G.: The Big Book of Concepts. MIT Press, Cambridge (2002)
6. Glanzberg, M.: Meaning, concepts, and the lexicon. Croatian J. Philos. **11**(1), 1–29 (2011)
7. OED Online: "Broadcast". Oxford University Press. http://www.oed.com
8. Hamilton, W.L., Leskovec, J., Jurafsky, D.: Diachronic word embeddings reveal statistical laws of semantic change. arXiv preprint arXiv:1605.09096 (2016)
9. Wevers, M., Kenter, T., Huijnen, P.: Concepts through time: tracing concepts in Dutch Newspaper Discourse (1890–1990) using word embeddings. In: Digital Humanities 2015, Sydney (2015)

10. Kenter, T., Wevers, M., Huijnen, P.: Ad hoc monitoring of vocabulary shifts over time. In: Proceeding of 24th ACM International on Conference on Information and Knowledge Management, pp. 1191–1200. ACM, New York (2015)
11. Wang, X., McCallum, A.: Topics over time: a Non-Markov Continuous-Time Model of topical trends. In: Proceedings of 12th ACM SIGKDD International Conference on Knowledge Discovery and Data Mining, pp. 424–433. ACM, New York (2006)
12. Blei, D.M., Lafferty, J.D.: Dynamic topic models. In: Proceeding of 23rd International Conference on Machine Learning, pp. 113–120 (2006)
13. Hall, D., Jurafsky, D., Manning, C.D.: Studying the history of ideas using topic models. In: Proceedings, Conference on Empirical Methods on Natural Language Processing (EMNLP), pp. 363–371. Association for Computational Linguistics, East Stroudsburg, Pennsylvania (2008)
14. Sigrist, R., Rawat, V.: Topic evolution in a stream of documents. In: Proceedings of SIAM International Conference on Data Mining. SIAM, Philadelphia (2009)
15. Gulordava, K., Baroni, M.: A distributional similarity approach to the detection of semantic change in the Google Books N-gram corpus. In: Proceedings of the EMNLP 2011 Geometrical Models for Natural Language Semantics (GEMS) Workshop. Association for Computational Linguistics, East Stroudsburg, Pennsylvania (2011)
16. Wijaya, D.T., Yeniterzi, R. Understanding semantic change of words over centuries. In: Proceeding DETECT, International Workshop on DETecting and Exploiting Cultural diversiTy on the Social Web, pp. 35–40. ACM, New York (2011)
17. Pesquita, C., Couto, F.M.: Predicting the extension of biomedical ontologies. PLoS Comput. Biol. **8**(9), e1002630 (2012)
18. Early English Books Online. Web, 02 October 2017. https://eebo.chadwyck.com/search?SCREEN=search_advanced.htx
19. OED Online: "Biology". Oxford University Press. http://www.oed.com
20. Recchia, G., Jones, E., Nulty, P., Regan, J., de Bolla, P.: Tracing shifting conceptual vocabularies through time. In: Proceeding of Detection, Representation and Management of Concept Drift in Linked Open Data (Drift-a-LOD), Bologna, Italy, 20 November 2016, pp. 2–9 (2016). CEUR-WS.org/Vol-1799/Drift-a-LOD2016_paper_1.pdf

The SemaDrift Protégé Plugin to Measure Semantic Drift in Ontologies: Lessons Learned

Thanos G. Stavropoulos[✉], Stelios Andreadis, Efstratios Kontopoulos, Marina Riga, Panagiotis Mitzias, and Ioannis Kompatsiaris

Centre for Research & Technology Hellas, 6th Km Charilaou – Thermi,
57001 Thessaloniki, Greece
{athstavr,andreadisst,skontopo,mriga,pmitzias,ikom}@iti.gr

Abstract. Semantic drift is an active research field, which aims to identify and measure changes in ontologies across time and versions. Yet, only few practical methods have emerged that are directly applicable to Semantic Web constructs, while the lack of relevant applications and tools is even greater. This paper presents the findings, current limitations and lessons learned throughout the development and the application of a novel software tool, developed in the context of the PERICLES FP7 project, which integrates currently investigated methods, such as text and structural similarity, into the popular ontology authoring platform, Protégé. The graphical user interface provides knowledge engineers and domain experts with access to methods and results without prior programming knowledge. Its applicability and usefulness are validated through two proof-of-concept scenarios in the domains of Web Services and Digital Preservation; especially the latter is a field where such long-term insights are crucial.

Keywords: Semantic drift · Concept drift · Semantic change · Ontologies · Protégé

1 Introduction

Evolving semantics, also referred to as *semantic change*, is an active and growing area of research that observes and measures the phenomenon of change in the meaning of concepts within knowledge representation models, along with their potential replacement by other meanings over time. In the Semantic Web (also known as Web 3.0), the representation of the underlying knowledge is typically assumed by ontologies. Thus, it can be easily perceived that semantic change can have drastic consequences on the use of ontologies in Semantic Web and Linked Data applications. In this setting, semantic change, i.e. the structural difference of the same concept in two ontologies [1], relates to various lines of research. Such examples are *concept* and *topic shift* [2], *concept change* [3], *semantic decay* [4], *ontology versioning* [5] and *evolution* [6]. A brief disambiguation of these terms can be found in [7].

This paper focuses on *semantic drift*, i.e. the phenomenon of ontology concepts gradually changing as knowledge evolves, obtaining possibly different meanings, as interpreted by various user communities or in a different context, risking their rhetorical,

P. Ciancarini et al. (Eds.): EKAW 2016 Satellite Events, LNAI 10180, pp. 29–39, 2017.
DOI: 10.1007/978-3-319-58694-6_3

descriptive and applicative power [8]. *Concept drift* can refer to this language-related phenomenon, but also to abrupt parameter value changes in data mining [9].

The SemaDrift plugin for the Protégé platform[1] aims at assisting a wider audience to monitor and manage semantic drift. The plugin was developed in the context of the PERICLES FP7 project[2], integrating and extending existing studies [2] and previously developed open, reusable methods [7]. A graphical user interface (GUI) makes the tool more attractive for a wider audience, including non-experts, towards accessing methods for monitoring evolving semantics, as a vehicle to measure and manage ontology change. The tool is validated through two realistic real-world applications, in Digital Preservation and Web Services, demonstrating its applicability and usefulness. This paper complements the contributions of [10], where the plugin was initially presented, with valuable insights and details regarding the metrics used and developed, discussion and limitations and extended future work.

The rest of the paper is structured as follows: Sect. 2 presents related work in metrics and tools for measuring drift. Section 3 presents the proposed framework consisting of the drift metrics and the tools functionality. Section 4 presents proof-of-concept applications. Section 5 discusses insights with respect to current limitations, while conclusions and future work are listed in the final section.

2   Related Work

Measures of semantic richness of Linked Data concepts have been investigated in [4], proving that increasing reuse of concepts decreases its semantic richness. Other studies have examined change detection between two ontologies at a structural or content level [1]. Concept drift has been measured either by clustering while populating ontologies [11] or by applying linguistic techniques on textual concept descriptions [12]. A vector space model by random indexing has been utilized to track changes of an evolving text collection [8]. A strategy to represent change has been based on ontology evolution [6]. However, most of these techniques are not directly applicable to Semantic Web constructs or present limited statistical data.

An appealing solution transfers the notions of *label*, *extension* and *intension* from machine learning concept drift to semantic drift, further defining them in ontology terms [2]. Much philosophical debate examines how and by which properties a concept can be identified across time and appropriate formalization [13]. Some have utilized the notions of perdurance and endurance [14], so as to seek identity, by defining rigid properties that have to be persistent across instances and, thus, can identify entities [9]. Further works have followed, focusing on the extensional drift aspect of statistical data [15]. In this work we adopt, implement and integrate the methods in [2] into a familiar application for knowledge engineers, targeting not only the lack of reproducible cross-domain metrics for semantic drift, but also the lack of similar graphical user interfaces.

---

[1] The Protégé Ontology Editor: http://protege.stanford.edu.
[2] PERICLES FP7 project: www.pericles-project.eu.

3   The SemaDrift Protégé Plugin

SemaDriftaims to bring novel semantic drift measuring capabilities into a popular ontology development platform, Protégé. Protégé offers many advantages to be chosen as the tool to integrate with. Traditionally as a desktop application, and recently also as a web application[3], it provides a user-friendly graphical interface for authoring ontologies and included entities, and naturally constitutes a more flexible alternative to plain text or RDF/OWL, especially for the unfamiliarized users.

Additionally, Protégé integrates a variety of add-onsdeveloped by its highly active community of users, like e.g. reasoners and third-party plugins, such as query tools and rich graph visualizations. The SemaDrift plugin fits perfectly into this multi-purpose environment, allowing users to interleave drift measurement, ontology authoring, reasoning, querying and visualization.

Both the plugin and its underlying drift metrics library are available online[4] under Apache V2 license. The metrics library was developed in Java and is based on the OWL API[5] for parsing ontologies and on Simmetrics[6] for implementing text similarity algorithms. The plugin is written in Java Swing[7], as required by Protégé.

3.1   Semantic Drift Metrics

The drift metrics presented here implement and extend previous work in the field of concept drift [7], where highly applicable notions and metrics for measuring concept drift in the context of data mining have successfully been transferred to semantic drift. The method to measure concept drift in semantics considers two basic factors: (a) the different aspects of change, and (b) whether concept identity is known or not. The aspects of change can be:

- *Label*, which refers to the description of a concept, via its name or title;
- *Intension*, which refers to the characteristics implied by it, via its properties;
- *Extension*, which refers to the set of things it extends to, via its number of instances.

Meanwhile, the correspondence of a concept across versions can be either known or unknown, resulting in two different approaches for measuring change:

- *Identity-based approach* (i.e. known concept identity): Assessing the extent of shift or stability of a concept's meaning is performed under the assumption that its identity is known across ontologies. For instance, considering an ontology *A*, and its evolution, ontology *B*, each concept of *A* is known to correspond to a single, known concept of *B*.

---

[3] https://webprotege.stanford.edu.

[4] SemaDrift Library API and Protégé Plugin online: http://mklab.iti.gr/project/semadrift-measure-semantic-drift-ontologies, hosted at MKLab tools: http://mklab.iti.gr/results/tools.

[5] OWL API: http://owlapi.sourceforge.net.

[6] https://github.com/Simmetrics/simmetrics.

[7] https://docs.oracle.com/javase/7/docs/api/javax/swing/package-summary.html.

- *Morphing-based approach* (i.e. unknown concept identity): Each concept is pertaining to just a single moment in time (ontology), while its identity is unknown across versions (ontologies), as it constantly evolves/morphs into new, even highly similar, concepts. Therefore, its change has to be measured in comparison to every concept of an evolved ontology.

The currently proposed method considers the more general morphing-based approach and considers drift as the dissimilarity of two maximally similar concepts in two versions [2]. Despite several methods have been proposed to seek identity correspondence across versions [9], they still can be domain or model dependent, mandating for ad-hoc expert knowledge in the form of annotations, user input or using explicit identities. In order to measure change, the meaning of each concept at a given point t (e.g. in time) is defined as a set of the three different aspects, as follows:

$$C^t = \; < label_t(C), int_t(C), ext_t(C) >$$

where C^t denotes the meaning of concept C at point t. Each of its aspects, $label_t(C)$, $int_t(C)$ for intensional and $ext_t(C)$ for extensional, is measured as follows:

$$label_t(C) = \{l, \mid \forall \langle C, rdfs\text{: } label, l \rangle \in T\}$$

$$int_t(C) = \{i \mid i = \langle C, p, x \rangle \vee i = \langle x, p, C \rangle, \; p = rdfs\text{:}domain \vee p$$
$$= rdfs\text{:}range, \forall i \in T\}$$

$$ext_t(C) = \{x \mid \forall \langle x, rdf\text{:}type, C \rangle \in T\}$$

where T is the set of all triples in the ontology version t. Namely (a) label is the *rdfs:label* of a concept (a string), (b) intension is a set of triples (i.e. the properties that involve the concept, calculated as the union of all RDF triples with C in the subject or object position of OWL Object Properties or OWL Datatype Properties) and (c) extension is the set of strings (i.e. the names of instances with the concept as value of *rdf:type*). Due to the morphing based approach, each concept's drift is measured as the average drift to all concepts of the next ontology. Comparisons for strings are made using the Monge-Elkan algorithm [16], found to optimally suit strings in ontologies such as CamelCase or snake_case, and Jaccard similarity for sets.

In detail, if n_2 is the total number of concepts in t_2, we define label, intensional and extensional drifts of C between versions t_1 and t_2 as follows:

$$label_{t_1 \to t_2}(C) = \frac{\sum_{i=1}^{n_2} MongeElkan\left(label_{t_1}(C), label_{t_2}(C_i)\right)}{n_2}$$

$$int_{t_1 \to t_2}(C) = \frac{\sum_{i=1}^{n_2} Jaccard\left(int_{t_1}(C), int_{t_2}(C_i)\right)}{n_2}$$

$$ext_{t_1 \to t_2}(C) = \frac{\sum_{i=1}^{n_2} Jaccard\left(ext_{t_1}(C), ext_{t_2}(C_i)\right)}{n_2}$$

A holistic aspect, *whole* is defined as their average:

$$whole_{t_1 \to t_2}(C) = \frac{label_{t_1 \to t_2}(C) + int_{t_1 \to t_2}(C) + ext_{t_1 \to t_2}(C)}{3}$$

3.2 Functionality

A comprehensive look at the SemaDrift plugin functionality is shown in Fig. 1. The tool provides a subset of the basic functions of the underlying SemaDrift API, in a graphical manner. For that purpose, it exposes some of its functions and accommodates the outcomes in suitable user controls using the Java Swing library. This edition of the plugin focuses on ontology pairs, i.e. two versions of the same ontology, in order to provide more insight into them and their differences, fitting also into the Protégé workspace philosophy. Usually, the users work on a single ontology at a time, which is always displayed as a tree hierarchy of classes at the left pane. Then, plugins occupy the right pane, which is free to accommodate their functions (Fig. 1).

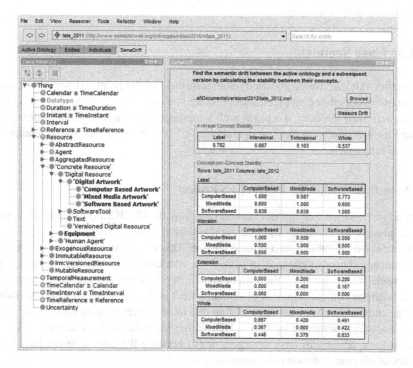

Fig. 1. SemaDrift Protégé plugin: The native tree hierarchy of the open ontology is shown on the left, while the plugin-provided content resides on the right, showing a second ontology to compare to, accompanied by the respective measurements.

As a first step the user has to select the pair of ontologies for which to measure drift. To take advantage of the environment, the plugin assumes that the first selected ontology is the one currently loaded in Protégé, allowing also its in-depth visualization, reasoning and query execution. The second ontology can be selected from the SemaDrift pane using the "Browse" button to look through local or remote storage.

After both ontologies are available, pressing on the *"Measure Drift"* button will display the SemaDrift metric results. Stability, as a measure of drift, is shown in two sections: overall average stability per aspect and concept pair stability for all aspects. The first section constitutes the most generic, abstract measure of drift. It displays a table with the average drift of all concepts from the former ontology to the latter, per each of the four aspects: label, intension, extension and whole. Naturally, the measurements are derived using the metrics and algorithms for each aspect described in the previous section, yielding a value from zero (no similarity) to one (full similarity).

The second section of results is displayed in respective tables. Each table row corresponds to a concept of the former ontology and each column to a concept of the latter. Consequently, each cell holds the similarity metric (i.e. concept stability) between each pair of concepts. These similarity values between pairs can further be utilized by users for different purposes; examples are given in the next section.

Concluding, the GUI in its current form is in essence a first step towards measuring semantic drift in a graphical manner. Its many possible extensions considered are given in the final section of this paper.

4   Use Case Scenarios

This section validates the applicability and usefulness of the proposed SemaDrift tools through two proof-of-concept scenarios presented below in the domains of Digital Preservation and Web Services. To begin with we present the framework for building the tool and its special ties to the first domain, digital preservation. The two domains are presented afterwards.

4.1   SemaDrift Within the PERICLES Project

As already mentioned, SemaDrift is an outcome from the PERICLES FP7 project, which aims to address the challenge of ensuring that digital content remains accessible in an environment that is subject to continual change. This can encompass not only technological change, but also changes in semantics, academic or professional practice, or society itself, which can affect the attitudes and interests of the various stakeholders that interact with the content. The risk of semantic drift is that, as a fallout from inevitable language change that has been accelerating due to an interplay of factors, future users may lose access to semantic content in DOs, either (a) because the concepts and/or the words as their labels will have changed, or (b) because the same concept may have different labels over separate user communities.

Thus, after thoroughly investigating the terminology and pertinent concepts, we selected a key methodology based on the notion of ontology evolution constituting the underlying intellectual framework of our efforts to address evolving semantics. Our primary aim was to anchor our plans to progress in real user needs proposed by project partners, link them to existing approaches and design appropriate experiments to base the solutions on partner data; this is exactly the case with the Digital Preservation scenario presented in following subsection.

4.2 Semantic Drift in Digital Preservation

The field of Digital Preservation shows much need for change detection across time and versions. The realistic scenario presented here serves as a means for validating the applicability of the framework in real-world conditions, while showcasing the usability of the SemaDrift Protégé plugin.

For the scenario a dataset was synthesized based on a ten-year period (2003–2013) acquisition log of software-based artworks by Tate Galleries, London[8]. A set of ontologies were developed, one for each year in the decade, modelling the respective domain concepts based on the Software-Based Art ontology found in [17, 18]. A key problem we wanted to address was to investigate whether the terms used for indexing the artworks (i.e. *"Computer-based Art"*, *"Mixed-Media Art"* and *"Software-based Art"*) refer to semantically similar or different notions. Thus, the ontologies of the dataset were loaded in SemaDrift and the outcomes of the proposed methods were visualized, yielding otherwise inaccessible insights about semantic drift across time.

Each pair of ontologies, either temporally consecutive or not, can be loaded and examined in SemaDrift. After examining all pairs in our scenario, here we show case for simplicity only three concepts from the 2011 and 2012 versions on Fig. 1. The minimum stability is noted in the extensional aspect, by its low average concept stability. Investigating further in concept-per-concept Stability, instances of *ComputerBased*art in 2011 are shared between *MixedMedia* and *SoftwareBased* in 2012, while some *MixedMedia* instances are now categorized as *SoftwareBased*.

The other aspects are in fact stable, bearing high values. Labels are unchanged across the matrix diagonal. The other values actually represent cross-concept similarity, which can be misinterpreted as drift; an issue that future identity-based methods can tackle. The same holds for intension: properties are retained across versions, but also all three concepts are similar, as they share half of their properties (yielding 0.5 cross-concept similarity and an overall 0.667 average).

All in all, after inspecting the results together with the domain experts from Tate, we concluded that the proposed tool and methodology indeed capture the underlying terminology change, as certain results coincided with official Tate policies (e.g. total abandonment of using the term "Computer-based" after 2012).

[8] Partnership with TATE within the context of the PERICLES FP7 project provided realistic knowledge for the generated models.

4.3 Semantic Drift in the Web Services Domain

The second scenario uses versions 1.0 and 1.2 of the OWL-S ontology[9] for semantic markup of Web Services, in order to demonstrate the tool's scalability as well as its ability to quickly pinpoint semantic drifts in ontologies. In OWL-S, each service has a *Profile*, a *Grounding* and a *Process Model*. A critical piece of *Profile* metadata are operation IOPEs, defining its Input and Output information (e.g. credit card number and total price), Preconditions required to proceed with it (e.g. credit card clearance) and its Effects (e.g. transferring ownership of goods or granting access). In this scenario, the *Profile* ontology changes are immediately apparent in the SemaDrift plugin.

As Fig. 2 shows, average concept drift originates from intension, which is investigated further. No instances exist for measuring extension, and labels changed only slightly. Some concepts vanished (e.g. *ConditionalEffect*, *ServiceCategory*) and some stayed the same (symmetrical concepts *Process*, *Parameter*). However, changes were detected in *Profile*, which bears altered properties and *Precondition*, which migrated to *Condition*. Other concepts present full stability simply because they bear no properties (marked as gray, while the remaining non-zero entries are marked in yellow).

Average Concept Stability

Label	Intensional	Extensional	Whole
0.807	0.139	1.000	0.648

Concept-per-Concept Stability
Rows: Profile 1.0 Columns: Profile 1.2
Intension

	Condition	Input	Output	Parameter	Process	Profile	Result	ServiceProfile
ConditionalEffect	0.000	0.000	0.000	0.000	0.000	0.000	0.000	0.000
ConditionalOutput	0.000	1.000	1.000	0.000	0.000	0.000	0.000	1.000
Input	0.000	1.000	1.000	0.000	0.000	0.000	0.000	1.000
Parameter	0.000	0.000	0.000	1.000	0.000	0.000	0.000	0.000
Precondition	1.000	0.000	0.000	0.000	0.000	0.000	0.000	0.000
Process	0.000	0.000	0.000	0.000	1.000	0.000	0.000	0.000
Profile	0.000	0.000	0.000	0.000	0.000	0.200	0.000	0.000
ServiceCategory	0.000	0.000	0.000	0.000	0.000	0.000	0.000	0.000

Fig. 2. Average concept and concept-per-concept intension stability for OWL-S, 1.0 vs 1.2. (Color figure online)

5 Discussion

SemaDrift is being currently used in various settings in order to improve its efficiency. Moreover, the tool is currently being evaluated by participants in the PERICLES-funded PhD course on the "Dynamics of Knowledge Organization"[10], involving students from around Europe. Besides the obvious limitations inherent to the GUI, the experience so far revealed also restrictions and dead-ends related to the metrics and methods. This section discusses these findings in detail.

A prominent issue inherent to the metrics is high similarity falsely perceived as high concept stability. A concept may appear as a highly stable one, yielding high stability

---

[9] OWL-S ontology: https://www.w3.org/Submission/OWL-S/.
[10] https://goo.gl/q9AwSR.

metrics. But actually, many times, those high metrics are due to high similarity to all concepts. Metrics susceptible to that flaw are text-similarity metrics in the label method. This feature turns out to be pertinent to the morphing-based method, as it does not really select a corresponding concept, increasing stability for as many as the similar concepts are. However, in more realistic situations this uncertainty lifts as the ontology grows in size, rendering similar names more and more unlikely. Another approach would be to completely erase this uncertainty by using an identity approach, at the cost of manual labour to annotate the corresponding concepts and the GUI means to do that.

Another issue discovered through the use cases is the tool and the metrics treating two empty sets as similar. While this is naturally true, it can be misleading while studying drift. In other words, it leads to the user thinking two fairly unrelated concepts to be highly similar (1.0) only because their respective property/instance sets are empty. While this was never before considered by any drift metrics so far, it was highly visible through the tool applied in the scenario. The issue essentially boils down to discriminating (and GUI highlighting) the total matches (1.0) that are in face empty sets versus the ones that are non-empty, as done manually on Fig. 2. All the above limitations are further considered and plans to resolve them are laid out in the next section.

6   Conclusions and Future Work

This paper presented the metrics, development, use cases and lessons learned for a Protégé plugin for measuring semantic change in terms of concept drift. Based on state-of-the-art notions, methods for measuring label, intensional, extensional and whole (total) drift have been adapted, optimized and implemented in the SemaDrift open source software library. The proposed domain-independent, cross-platform software tool was integrated with the popular Protégé platform, enriching its multi-purpose knowledge engineering environment with semantic drift measurement capabilities, as showcased through two proof-of-concept scenarios, in Digital Preservation and semantic markup for Web Services.

SemaDrift shows much room for future improvement, both as a GUI and in terms of the metrics and methods. Initially, as discussed above, the tool could aid to lift certain drawbacks in the method. Namely, discriminating the identities of concepts across versions could eliminate the error in showing concepts with incidentally similar labels to others as highly stable ones. While now the method does not require further input to pinpoint identities, users could do so in the future, yielding a series of identity-based metrics which could be more valuable in certain cases. Additionally, having the tool check at GUI-level which exact matches of concept property or instance sets (rated 1.0) refer to empty or non-empty sets; in the former case, the cells can be highlighted as incidental matches while in the latter as really interesting cases of migrating properties or instances.

Regarding the GUI, the chain of ontology versions to compare to, which is currently only limited to two, will be increased using more GUI controls. Combined with visualization capabilities, the user will be able to view entire morphing chains effortlessly, targeting long-term investigation. Finally, a standalone desktop application is already underway to allow this level of flexibility at the GUI level as well as to appeal to a wider audience, which is then planned to evaluate the tool for user-friendliness and effectiveness for the task

Acknowledgements. This research received funding by the European Commission Seventh Framework Programme under Grant Agreement Number FP7-601138 PERICLES.

References

1. Tury, M., Bieliková, M.: An approach to detection ontology changes. In: Workshop Proceedings of the Sixth International Conference on Web Engineering - ICWE 2006, p. 14. ACM (2006)
2. Wang, S., Schlobach, S., Klein, M.: Concept drift and how to identify it. J. Web Semant. **9**, 247–265 (2011)
3. Uschold, M.: Creating, integrating and maintaining local and global ontologies. In: Proceedings of the First Workshop on Ontology Learning (OL-2000) in Conjunction with the 14th European Conference on Artificial Intelligence (ECAI 2000), Berling, Germany
4. Pareti, P., Klein, E., Barker, A.: A Linked data scalability challenge: concept reuse leads to semantic decay. In: Proceedings of the ACM Web Science Conference. ACM Press-Association for Computing Machinery (2016)
5. Yildiz, B.: Ontology evolution and versioning. Technical report, TU Vienna (2006)
6. Stojanovic, L., Maedche, A., Motik, B., Stojanovic, N.: User-driven ontology evolution management. In: Knowledge Engineering and Knowledge Management: Ontologies and the Semantic Web, pp. 133–140 (2002)
7. Stavropoulos, T.G., Andreadis, S., Riga, M., Kontopoulos, E., Mitzias, P., Kompatsiaris, I.: A framework for measuring semantic drift in ontologies. In: 1st International Workshop on Semantic Change and Evolving Semantics (SuCCESS 2016). CEUR Workshop Proceedings, Leipzig, Germany (2016)
8. Wittek, P., Darányi, S., Kontopoulos, E., Moysiadis, T., Kompatsiaris, I.: Monitoring term drift based on semantic consistency in an evolving vector field. In: Proceedings of the International Joint Conference on Neural Networks, pp. 1–8. IEEE (2015)
9. Meroño-Peñuela, A., Hoekstra, R.: What is linked historical data? In: Janowicz, K., Schlobach, S., Lambrix, P., Hyvönen, E. (eds.) EKAW 2014. LNCS, vol. 8876, pp. 282–287. Springer, Cham (2014). doi:10.1007/978-3-319-13704-9_22
10. Stavropoulos, T.G., Andreadis, S., Kontopoulos, E., Riga, M., Mitzias, P., Kompatsiaris, I.: SemaDrift: a Protégé plugin for measuring semantic drift in ontologies. In: Hollink, L., Darányi, S., Meroño Peñuela, A., Kontopoulos, E. (eds.) 1st International Workshop on Detection, Representation and Management of Concept Drift in Linked Open Data (Drift-a-LOD) in Conjunction with the 20th International Conference on Knowledge Engineering and Knowledge Management (EKAW). CEUR Workshop Proceedings, Bologna, Italy, vol. 1799 (2016)
11. Fanizzi, N., d'Amato, C., Esposito, F.: Conceptual clustering and its application to concept drift and novelty detection. In: Bechhofer, S., Hauswirth, M., Hoffmann, J., Koubarakis, M. (eds.) ESWC 2008. LNCS, vol. 5021, pp. 318–332. Springer, Heidelberg (2008). doi: 10.1007/978-3-540-68234-9_25
12. Gulla, J., Solskinnsbakk, G., Myrseth, P.: Semantic drift in ontologies. In: Proceedings of 6th International Conference on Web Information Systems and Technologies (WEBIST), Valencia, Spain, pp. 13–20 (2010)

13. Guarino, N., Welty, C.: A formal ontology of properties. In: Dieng, R., Corby, O. (eds.) EKAW 2000. LNCS, vol. 1937, pp. 97–112. Springer, Heidelberg (2000). doi: 10.1007/3-540-39967-4_8

14. Gangemi, A., Guarino, N., Masolo, C., Oltramari, A., Schneider, L.: Sweetening ontologies with DOLCE. In: Gómez-Pérez, A., Benjamins, V.Richard (eds.) EKAW 2002. LNCS, vol. 2473, pp. 166–181. Springer, Heidelberg (2002). doi:10.1007/3-540-45810-7_18

15. Mitzias, P., Riga, M., Waddington, S., Kontopoulos, E., Meditskos, G., Laurenson, P., Kompatsiaris, I.: An ontology design pattern for digital video. In: Proceedings of the 6th Workshop on Ontology and Semantic Web Patterns (WOP 2015) (2015)

16. Monge, A.E., Elkan, C.: The field matching problem: algorithms and applications. In: 2nd International Conference on Knowledge Discovery and Data Mining (KDD), pp. 267–270 (1996)

17. Lagos, N., Kontopoulos, E., Riga, M., Mitzias, P., Meditskos, G., Waddington, S., Laurenson, P.: Designing for inconsistency–the dependency-based PERICLES approach. In: East European Conference on Advances in Databases and Information Systems, pp. 458–467. Springer International Publishing, Cham (2015)

18. Kontopoulos, E., Riga, M., Mitzias, P., Andreadis, S., Stavropoulos, T., Lagos, N., Vion-Dury, J.-Y., Meditskos, G., Falcão, P., Laurenson, P., Kompatsiaris, I.: Ontology-based representation of context of use in digital preservation. In: 1st Workshop on Humanities in the Semantic Web - WHiSe, Co-located with the 13th Extended Semantic Web Conference (ESWC 2016), Heraklion, Crete, Greece. CEUR Workshop Proceedings, vol. 1608, pp. 65–72 (2016)

Combining Distributional Semantics and Structured Data to Study Lexical Change

Astrid van Aggelen[✉], Laura Hollink, and Jacco van Ossenbruggen

Centrum Wiskunde & Informatica, Amsterdam, The Netherlands
a.e.van.aggelen@cwi.nl

Abstract. Statistical Natural Language Processing (NLP) techniques allow to quantify lexical semantic change using large text corpora. Word-level results of these methods can be hard to analyse in the context of sets of semantically or linguistically related words. On the other hand, structured knowledge sources represent semantic relationships explicitly, but ignore the problem of semantic change. We aim to address these limitations by combining the statistical and symbolic approach: we enrich WordNet, a structured lexical database, with quantitative lexical change scores provided by HistWords, a dataset produced by distributional NLP methods. We publish the result as Linked Open Data and demonstrate how queries on the combined dataset can provide new insights.

Keywords: Lexical semantics · NLP · Knowledge bases · Linked Open Data

1 Introduction

How words have been used in discourse over time, have adopted new senses or changed their meaning is studied in the humanities and social sciences (e.g., [1–3]) and information sciences (e.g., [4,6]). We make a case for interlinking structured knowledge bases with the outcomes of Natural Language Processing (NLP) methods for the purpose of tracing language change over time.

Semantic change in words is increasingly modelled using distributional NLP methods (word embeddings) (e.g. [8,9]). These techniques represent the meaning of a word in terms of its tendency to co-occur with other words in the lexicon, as observed in large text corpora. Since this results in vectors, cosine distances can be used to quantify the correspondence between two such representations. When vectors are assembled for the lexicon in separate time spans, the notion of distance can be applied to find a word's nearest neighbours within a time frame, or to calculate the degree of change a word underwent from one time interval to the next.

However, word embeddings alone are not sufficient to gain insight into the dynamics of the lexicon and to elicit follow-up questions or hypotheses. They

---

This paper is an extended version of [13].

© Springer International Publishing AG 2017
P. Ciancarini et al. (Eds.): EKAW 2016 Satellite Events, LNAI 10180, pp. 40–49, 2017.
DOI: 10.1007/978-3-319-58694-6_4

operate on the level of individual terms, often without metadata, making it hard to see patterns and connections. It is thinkable, though, that language change affects not just individual terms but also clusters of (related) terms, that show interaction in their motions of semantic drift. Also, some types of words might change more than others. Structured knowledge sources can help derive such insights. For instance, lexical resources allow to group and connect findings for individual terms by their relation.

Conversely, statistical findings of lexical change could provide a useful addition to structured knowledge bases, as these typically contain only static, contemporary facts. One example application is in annotating historic documents, where the terms might have changed their meaning and are difficult to map onto metadata instances. Khan et al. [7] have introduced a vocabulary, LemonDIA, to express qualitative (linguistic) typifications of lexical shifts. This vocabulary is compatible with, and the knowledge it expresses is complementary to, the data curated in this project.

This paper is a step towards the goal of a structured, interconnected knowledge source of diachronic lexical semantics. It presents an interlinking effort between HistWords, a unique corpus of (open) lexical change data, and Word-Net, a lexical database which is part of the Linked Open Data cloud. This combination results in a knowledge graph were concepts, linguistic data elements such as lexemes, and semantic change scores can be queried together. By publishing the data in the Resource Description Framework, we aim to contribute to the (re-)usability of these open corpora.

In the remainder of this paper, we discuss how the HistWords data were linked to lexical entries in WordNet and how the result was represented in an RDF data model. Example queries on this aggregated dataset demonstrate the use as well as the limitations of the approach.

2   Source Data

HistWords. HistWords is a research project of *Word embeddings for Historical Text* at Stanford University that has produced sets of word embeddings and cross-decade lexical change scores. We used all ready-made lexical change scores for English[1], i.e., for the 10.000 most frequent (averaged over decades) words from the English Google N-Grams dataset[2] excluding proper nouns. The entries in this dataset are not lemmatised, disambiguated or part-of-speech tagged, hence each similarity score reflects all senses and grammatical functions in which the word can occur. The linking effort to WordNet, which does distinguish between different parts of speech, does not solve this issue; rather, it makes it more explicit, as one can query for all possible lexical entries of different parts of speech that correspond to a given word, and for all of the word's senses. The similarity scores are given between discrete decades. They were calculated

---

[1] http://snap.stanford.edu/historical_embeddings/eng-all_sgns.zip, fullstats.
[2] http://storage.googleapis.com/books/ngrams/books/datasetsv2.html.

as the cosine similarity between the vector for a term derived from corpus material in one decade, and the vector for the same term derived from materials from the other decade. The embeddings were obtained by the word2vec skip-gram method with negative sampling [11] for each decade separately, followed by a transformation to project them into a single space; see [5] for details.

Figures are available for every two consecutive decades between 1810 and 2000; i.e., the degree of semantic stability of a lexical term from the 1810s to the 1820s, the 1820s to 1830s, and so on, up to 1980s–1990s. As an example, the word *gay* seems to have underwent semantic change between the 1980s and 1990s, where the cosine similarity between the two term representations fell to 0.91 (from 0.96 for the 1970s–1980s). In addition, there are figures for every decade vs. the 1990s, i.e., for 1810s vs. 1990s until 1980s vs. 1990s. These can be used to express the overall change of a lexeme in, for instance, the 20th century (1900s–1990s), or over the entire dataset (1810s–1990s). Due to corpus characteristics, some entries have (some) missing values, which were left out.

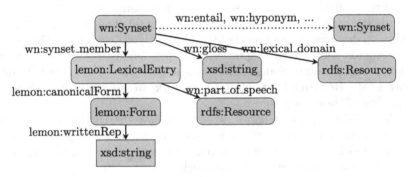

Fig. 1. The basic types of the WordNet RDF model. Prefix **wn** stands for the WordNet vocabulary.

WordNet. WordNet [12] is a lexical database of English. It is based on the idea of synsets, synonymous terms of a given grammatical category that express the same concept. One term hence can appear in multiple synsets; e.g., *gay*(adj.) is part of a synset of adjectives to denote *homosexual or arousing homosexual desires* (alongside *homophile* and *queer*) and a synset of adjectives for *bright and pleasant; promoting a feeling of cheer* (alongside *cheery* and *sunny*).

The RDF conversion of WordNet [10,14] (henceforth *RDF-WordNet*) used in this project is based on the Lemon vocabulary of linguistic annotations, completed with some WordNet-specific concepts. The basic resource types in RDF-WordNet are shown in Fig. 1. A `lemon:lexicalEntry` represents a single lemma of some grammatical type, of which RDF-WordNet counts 158K. The unique base form of each lemma (of type `Lemon:Form`) is pointed to by `lemon:canonicalForm`; inflectional variants (lexemes, word forms) are listed (by `lemon:otherForm`), though only for a minority of terms. The grammatical type is indicated through property `wn:part_of_speech`.

A `lemon:LexicalEntry` instance connects to one or more senses (`wn:Synset`) through `wn:synset_member`. Property `wn:gloss` relates a `wn:Synset` instance to its definition. When applicable, synsets are interrelated through semantic relations such as hyponymy, entailment, and meronymy. Additionally, each synset is categorised (using `wn:lexical_domain`) into one of 45 semantic-grammatical types such as `noun.artifact` and `verb.emotion`.

3   Approach

The sourced similarity scores were transformed into change data and connected to WordNet through (stemming and) string matching. The result was represented in RDF and OWL and made available as a Turtle download[3].

Deriving semantic change scores. The scores were converted to distance measures as we care about the degree of change more than the degree of stability of the words' meaning. This was done with an arc-cosine transformation rather than by the formula $1 - cosine\_similarity$ to stretch the scale of the change interval and trace more fine-grained differences. The semantic change rate thus lies between 0 and $\pi/2$ (in our dataset, between 0.09 and 1.48). For instance, between the 1980s and 1990s the change values ranged from 0.11 (*pepper*) to 1.12 (*web*). The rates for a larger period are generally higher than those for consecutive decades, e.g. 0.97 for *gang* between the 1810s and 1990s. The change scores have no clear absolute meaning but can be used contrastively between terms or time frames.

Linking HistWords to WordNet. The words in HistWords were mapped onto `lemon:LexicalEntry` instances in RDF-WordNet. First, we merged on an exact match between a word in HistWords and the value of the `lemon:writtenRep` property of the `lemon:Form` corresponding to the `lemon:LexicalEntry` instance. Since the HistWord words are not part-of-speech specific, they were mapped onto all lexical matches in WordNet, irrespective of grammatical type. This string matching step resulted in 7.365 matches for the 10.000 source words, mapped onto 10.956 `lemon:LexicalEntry` instances.

Aimed at representing as much of the source data as possible, unmapped HistWords entries were Porter stemmed and re-matched based on an exact match of the stem and a WordNet entry. We included the matches as new `lemon:lexicalEntry` instances with their unstemmed form as the canonical form, and connected them to their WordNet `lemon:lexicalEntry` counterparts through the `lemon:lexicalVariant` property. This brought the total number of mappings to 8.878 out of 10.000 source entries, connected to 12.469 `lemon:LexicalEntry` instances. In future work, it is likely that more words can be matched by refining our stem-and-match technique.

Data model. The resulting data, i.e., the tuples {lexical entry, decade1, decade2, change value}, were represented in RDF. Existing vocabularies were used where

---

[3] www.github.com/aan680/SemanticChange.

possible; newly introduced classes and properties are recognisable by the `cwi` prefix. Figure 2 illustrates how a `lemon:LexicalEntry` was connected to a node of type `cwi:SemanticChange` for each data tuple with a value and an onset and offset decade. The latter two were modelled, in accordance with OWL-Time[4], as intervals with a start and an end date.

Following OWL-Time ensures interoperability and supports temporal reasoning, but complicates queries for the semantic change of a word between two specified decades. For this reason we introduced a shortcut property for each set of decades, which directly connects a `lemon:LexicalEntry` instance to the semantic change value. The property URI encodes the decades it contrasts, e.g., `cwi:semantic_change_1910s-1920s` leads to the change score between the 1910s and the 1920s.

Note that instead of at the `lemon:LexicalEntry` level, we could have linked the HistWords entries to the `lemon:Form` level, representing the lexeme. We decided against this since it would greatly complicate the queries that we anticipate at the `LexicalEntry` or `Synset` level. This approach would have yielded only 334 mappings to inflectional variants, part of which were among the many mappings made in the second mapping step.

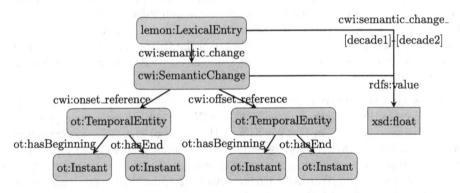

Fig. 2. A model for connecting WordNet entries to cross-decade scores of lexical change. Prefix `ot` stands for OWL-Time and `cwi` for the purpose-built vocabulary.

4   Usage Examples

We used the semantic web server ClioPatria [15] to query the RDF dataset of semantic change scores in combination with RDF-WordNet. Below we show example queries that exploit the connection to WordNet as a background source.

Example 1: average change per semantic/linguistic category. We collected the change rate between the decades 1810s and 1990s for all lexical entries as a proxy for their overall change score (alternatively, we could have averaged over all subsequent-decade scores), and related these scores to, first, their part of

---

[4] http://www.w3.org/TR/owl-time/.

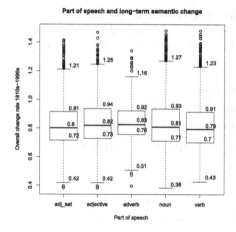

Part of speech and long-term semantic change

Fig. 3. The spread of the change score of lexical entries between the 1810s and 1990s by part of speech.

Table 1. The distribution over parts of speech of entries with a change score between 1810s and 1990s in our dataset and in RDF-WN.

POS	Dataset	%	RDF-WN	%
Noun	4410	44	118303	75
Verb	2021	20	11540	7
Adjective	1111	11	8358	5
Adjective satellite	1941	19	15068	10
Adverb	504	5	4475	3
TOTAL	9987	100	157744	100

speech property, and second, the WordNet domain they belong to. Recall that the HistWords index consists of raw word forms; thanks to WordNet, we can annotate these with grammatical and semantic information.

Figure 3 summarises the results and shows the spread of the change scores grouped by the parts of speech distinguished in WordNet. It shows that the change rates are evenly distributed over the grammatical categories. Looking at the distribution over parts of speech of the word entries themselves (Table 1), though, we see that our dataset contains relatively many verbs and adjectives and few nouns as compared to WordNet.

Table 2 shows examples of semantic domains and the mean change score of their lexical entries. Words in the given dataset that refer to processes, phenomena and events have seen a higher degree of change than words for food, feelings, or the weather. Note that besides by lexical domain, one can group findings by hypernymy relations between synsets. For instance, there is a synset of psychological states, with as its direct children synsets referring to depression, anxiety, irritation, nervousness, and more.

Example 2: the relationship between polysemy and semantic change. The synset structure of WordNet provides a simple way to quantify the degree of polysemy of a word. Hamilton et al. [5] find a positive correlation between the degree of change of words and their polysemy. They quantify polysemy using a co-occurrence network derived from a large text corpus, under the assumption that polysemous words tend to co-occur with words that do not tend to mutually co-occur. We were curious if we found the same effect when quantifying polysemy directly based on WordNet, as the number of senses (synsets) related to a word.

We plot the change score for 1810s–1990s of each word form (again, as a proxy for the overall change, as do [5]) against the number of synsets related

Table 2. Examples of WordNet semantic domains, ordered by the average change between the 1810s and 1990s (Change) of their lexical entries. Also given are the number of lexemes per domain (N) for which change scores are available and a few example words.

Domain	Change	N	Example words		
noun.process	0.91	117	Preservation	Infection	Decomposition
noun.linkdef	0.91	147	Proportion	Chemistry	West
adj.pert	0.91	147	Volcanic	Imperial	Legislative
...	...	...	...	...	...
noun.group	0.88	490	Flock	Bundle	Nine
noun.phenomenon	0.87	134	Moment	Mortification	Energy
noun.event	0.87	371	Expense	Vision	Climb
...	...	...	...	...	...
verb.emotion	0.80	188	Like	Triumph	Regret
noun.body	0.79	227	Wrist	Seat	Shoulder
noun.animal	0.79	159	Horn	Insect	Fish
noun.food	0.78	183	Game	Tea	Diet
noun.feeling	0.78	224	Gratitude	Gloom	Appetite
noun.plant	0.77	121	Tea	Foliage	Olive
verb.weather	0.74	36	Precipitate	Storm	Snow

to that word form (Fig. 4). One complicating factor is that a word form can be related to several lexical entries, for several parts of speech. Therefore, we also plot the change rate of lexical entries (rather than word forms) against their corresponding number of synsets. With neither of these tests, however, were we able to replicate the results of [5]: on our data we found just a very weak positive correlation (Kendall = 0.06 and 0.05 for words and lexical entries, respectively).

Example 3: exploring senses responsible for semantic drift. Upon browsing the dataset, we came across the word *yellow*. While this term did not display a great degree of change for most decades, we noticed a local peak in change for time period 1910s–1920s, where the score went from 0.25 (for 1900s–1910s) to 0.28 to then fall back to 0.23 (1920s–1930s) and climb up again to 0.25 (1930s–1940s). Clicking through to the senses of the word *yellow*, as RDF-WordNet allows one to do, we found a sense unknown to us. In addition to the colour, *yellow* is an adjective meaning *easily frightened*, with synonyms such as *chickenhearted*. Maybe the word was used in the two World Wars to refer to not-so-brave soldiers? This would explain the observed peaks. Since the change scores are not part-of-speech-, let alone sense-disambiguated, the answer is not in our dataset. For conclusions we would need to go back to the underlying (open source) text corpus, Google N-Grams, and have a close look at the term's occurrences. This example illustrates that our dataset is an addition to, not a substitute for, close reading methods.

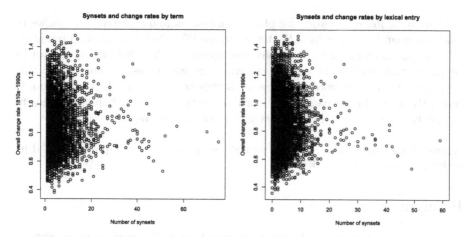

Fig. 4. Number of synsets and overall change rate by term (left) and by lexical entry (right).

5 Discussion and Future Work

This paper demonstrated how statistical findings of lexical semantic change can benefit from a connection to a structured knowledge base. Taking HistWords and WordNet as data sources, we have shown how this connection enables us to aggregate semantic change scores over semantic and linguistic categories.

We see various directions for future research. Firstly, there is the question of what type of change information is most valuable. The example queries on the dataset highlighted that lexical change scores, although useful, are heavily refined. Derived from word vectors, the distance figures no longer carry in them the distributions over vector components, which can be more telling about inter-word contrasts than a mere cosine measure. The word vectors in turn are derived from mentions in a text corpus, which are not included in the dataset itself. To check the findings against the source material, the user will need to query the Google N-Gram corpus. An open question remains what sort of data researchers in the field would like to see curated and integrated to benefit from a single source. A related question, which falls outside the scope of this project, is how to evaluate the change scores and draw reliable conclusions about lexical change.

Secondly, the dataset can be enriched in various ways. From the side of the NLP data curated in this project, nearest neighbour information could be added for words in time periods. Another addition we aspire is a score set based on part-of-speech-tagged words, such that the relation between the scores and the word senses are more clear-cut. From the side of knowledge bases, the change scores can be linked to additional sources. Examples are a cross-lingual dictionary such as BabelNet, to see if other languages display parallels in their lexical patterns of change, and a frame-semantic source like FrameNet as an alternative ground for grouping term-level findings.

Finally, as follow-up work, we plan to include more qualitative approaches to analyse semantic change data. We intend to use WordNet synsets as representing all concepts and meanings a word has ever referred to. By tracking a given word's similarity time series with each of the words in one synset, we hope to be able to assess whether the word has moved towards or away from the corresponding sense. By doing this for each of the word's synsets (senses) we hope to be able to automatically explicate the way in which a word has changed.

Acknowledgments. This work was partially supported by H2020 project VRE4EIC under grant agreement No. 676247.

References

1. Andreas Blank: Words and concepts in time: towards diachronic cognitive onomasiology (2003)
2. De Bolla, P.: The Architecture of Concepts: The Historical Formation of Human Rights. Oxford University Press, New York (2013)
3. Gabrielatos, C., Baker, P.: Fleeing, sneaking, flooding: a corpus analysis of discursive constructions of refugees and asylum seekers in the UK press, 1996–2005. J. Eng. Linguist. **36**(1), 5–38 (2008)
4. Gulordava, K., Baroni, M.: A distributional similarity approach to the detection of semantic change in the Google Books Ngram corpus. In: Proceedings of the GEMS 2011 Workshop on Geometrical Models of Natural Language Semantics, pp. 67–71. Association for Computational Linguistics (2011)
5. Hamilton, W.L., Leskovec, J., Jurafsky, D.: Diachronic word embeddings reveal statistical laws of semantic change. In: ACL 2016, pp. 1489–1501 (2016)
6. Kenter, T., Wevers, M., Huijnen, P., de Rijke, M.: Ad hoc monitoring of vocabulary shifts over time. In: Proceedings of the 24th ACM International Conference on Information and Knowledge Management, pp. 1191–1200. ACM (2015)
7. Khan, F., Díaz-Vera, J.E., Monachini, M.: Representing polysemy and diachronic lexico-semantic data on the Semantic Web. In: Proceedings of the Second International Workshop on Semantic Web for Scientific Heritage Co-located with 13th Extended Semantic Web Conference (ESWC 2016) (2016)
8. Kim, Y., Chiu, Y.-I., Hanaki, K., Hegde, D., Petrov, S.: Temporal analysis of language through neural language models. In: Proceedings of the ACL 2014 Workshop on Language Technologies and Computational Social Science, pp. 61–65 (2014)
9. Kulkarni, V., Al-Rfou, R., Perozzi, B., Skiena, S.: Statistically significant detection of linguistic change. In: Proceedings of the 24th International Conference on World Wide Web, pp. 625–635. ACM (2015)
10. McCrae, J.P., Fellbaum, C., Cimiano, P.: Publishing and linking WordNet using lemon and RDF. In: Proceedings of the 3rd Workshop on Linked Data in Linguistics (2014)
11. Mikolov, T., Dean, J.: Distributed representations of words and phrases and their compositionality. In: Advances in Neural Information Processing Systems (2013)
12. Miller, G.A.: WordNet: a lexical database for English. Commun. ACM **38**(11), 39–41 (1995)

13. van Aggelen, A., Hollink, L., van Ossenbruggen, J.: Combining distributional semantics and structured data to study lexical change. In: Proceedings of the 1st Workshop on Detection, Representation and Management of Concept Drift in Linked Open Data. CEUR Workshop Proceedings, vol. 1799, pp. 18–25. http://ceur-ws.org/Vol-1799

14. Van Assem, M., Gangemi, A., Schreiber, G.: Conversion of WordNet to a standard RDF/OWL representation. In: Proceedings of the Fifth International Conference on Language Resources and Evaluation (LREC 2006), Genoa, Italy, pp. 237–242 (2006)

15. Wielemaker, J., Beek, W., Hildebrand, M., van Ossenbruggen, J.: ClioPatria: A SWI-Prolog infrastructure for the semantic web. Semant. Web 7(5), 529–541 (2016). IOS Press

Second International Workshop on Educational Knowledge Management (EKM 2016)

Educational Knowledge Management (EKM 2016)

Inaya Lahoud[1]([⊠]), Nada Matta[2], Fouad Zablith[3],
and Sebastian Ventura[4]

[1] Galatasaray University, Istanbul, Turkey
clahoud@gsu.edu.tr
[2] University of Technology of Troyes, Troyes, France
Nada.matta@utt.fr
[3] American University of Beirut, Beirut, Lebanon
fzl3@aub.edu.lb
[4] University of Cordoba, Cordoba, Spain
sventura@uco.es

Abstract. The interest in knowledge engineering and knowledge management for the educational domain has been growing in recent years. This can be seen in the series of conferences organized by the International Educational Data Mining Society and in papers discussing the role of knowledge management in higher education. The EKM 2016 workshop aimed to bring together researchers, academic and professional leaders, consultants, and practitioners, from the domain of Semantic Web, data mining, machine learning, linked data, and natural language processing to discuss and share experiences in the educational area.

Knowledge management (KM) provides benefits to individual employees, to communities of practice, and to the organization itself. This three-tiered view of KM helps emphasize why KM is important today. It has been approved in different domains such as industrial and medical one. As education is increasingly occurring online or via educational software, resulting in an explosion of data, researchers started to explore this domain and new techniques are being developed and tested. They aim, for instance, to improve educational effectiveness, determine the key factors to the success of educational training, support basic research on learning, or manage educational training by satisfying the needs of a community, local industry, or professional development. The paper "Knowledge Management for Educational Information Systems: What Is the State of the Field?" written by Christopher A. Thorn in the journal "*Education Policy Analysis Archives*" emphasis on the need of KM techniques to face current evolution of educational systems especially the mobility of students/teachers involved in interactive learning. Thus, educational systems need KM to support and control the evolution of teaching and learning, and KM techniques must be adapted to education and take into account pedagogic dimensions.

The Second International Workshop on Educational Knowledge Management (EKM 2016) was organized as a satellite event to the 20th International Conference on Knowledge Engineering and Knowledge Management (EKAW), held in Bologna, Italy, November 19–23, 2016. The workshop was held on November 20.

In the call of papers, we invited submissions reporting original research related to any problem of managing and exploring information in the educational area in schools, colleges, universities, and other academic or professional learning institutions. A non-exhaustive list of topics for the workshop included the following:

- Knowledge management and ontology
- Knowledge management and multi-agents system
- Data mining and Semantic Web
- Knowledge acquisition, extraction, reuse
- Natural language processing to improve educational effectiveness
- Providing feedback to teachers and other stakeholders generated from EKM methods
- Generic frameworks, methods and approached for EKM
- Learner or student modeling
- Mining the results of educational research
- Educational process mining
- Improving educational software
- Evaluating teaching interventions
- Integrating data mining and pedagogical theory
- Integration pedagogical dimensions for knowledge appropriation
- Practice learning experiences and techniques
- Improving teacher support
- Semantic annotation
- Linked educational data
- Educational knowledge evolution
- Community of learning and community of practices

Two members of the Program Committee reviewed our submissions. The EKM Program Committee selected the best papers submitted to the workshop. Following the reviewers' recommendations, six full papers out of the 11 submissions were accepted for presentation and published as CEUR-WS proceedings volume 1780. The workshop was a half-day format with one invited speaker. Three selected best papers were extended, revised, and included in this Springer volume (EKAW 2016 proceedings).

As chairs of this workshop, we would like to thank everybody who was involved in the organization of EKM 2016, and we look forward to the next edition.

Program Committee

Ayako Hoshino	Data Science Research Laboratories, NEC Corp., Kawasaki, Japan
Catherine Faron Zucker	University of Nice Sophia Antipolis, France
Davide Taibi	Institute for Educational Technologies, Italian National Research Council, Italy
Marie-Hélène Abel	Sorbonne Universités, Université de technologie de Compiègne, France
Lisa Beinborn	TU Darmstadt, Germany
Kai Wing Chu	CCC Heep Woh College, Hong Kong, SAR China
Eddie Soulier	University of Technology of Troyes, France
Nitish Aggarwal	NUI Galway, Ireland
Lars Ahrenberg	Linköping University, Sweden
Christophe Paoli	Galatasaray University, Turkey

Learning Scorecard: Monitor and Foster Student Learning Through Gamification

Elsa Cardoso[1(✉)], Diogo Santos[2], Daniela Costa[2], Filipe Caçador[2],
António Antunes[2], and Rita Ramos[2]

[1] University Institute of Lisbon (ISCTE-IUL) and INESC-ID, Lisbon, Portugal
elsa.cardoso@iscte.pt
[2] University Institute of Lisbon (ISCTE-IUL), Lisbon, Portugal
{diogo_leo_santos, dscaal, filipe_cacador,
antonio_lorvao, rita_parada}@iscte.pt

Abstract. This paper presents the Learning Scorecard (LS), a platform that enables students to monitor their learning progress in a Higher Education course during the semester, generating the data that will also support the ongoing supervision of the class performance by the course coordinator. The LS uses gamification techniques to increase student engagement with the course. Business Intelligence best practices are also applied to provide an analytical environment for student and faculty to monitor course performance. This paper describes the initial design of the LS, based on a Balanced Scorecard approach, and the prototype version of the platform, currently in use by graduate and undergraduate students in the fall semester of 2016–2017.

Keywords: Balanced Scorecard · Business intelligence · Student learning · Gamification

1 Introduction

A recurrent problem in Higher Education is the lack of information about the progress of student learning in "real" time. Various statistics are calculated by Planning and Institutional Research offices offering a post analysis of academic success for each semester (e.g., evaluated, approved, and retention rates). Current pedagogic guidelines also encourage course coordinators to clearly define a set of tasks that students should perform autonomously additional to the course classes (e.g., exercises to be solved, basic and complementary bibliography that should be read). However, there is still little institutional support provided to students and faculty regarding the monitoring of the student learning experience and ongoing autonomous work completion throughout the semester. On the one hand, students do not know if they are correctly performing the proposed autonomous work, that is supposedly "a route to success in the course". On the other hand, a faculty has no information regarding the real commitment of students to the learning experience, apart from his/her experience-based perception of the class behavior.

© Springer International Publishing AG 2017
P. Ciancarini et al. (Eds.): EKAW 2016 Satellite Events, LNAI 10180, pp. 55–68, 2017.
DOI: 10.1007/978-3-319-58694-6_5

The Learning Scorecard (LS) is a platform that enables students to monitor their learning progress in a course during the semester, generating the data that will also support the ongoing supervision of the class performance by the course coordinator.

The LS was initially developed by a group of students in the context of a course on Decision Support Systems (DSS) of the master program in Computer Science Engineering in the 2015–2016 spring semester, at ISCTE – University Institute of Lisbon, a public University in Lisbon, Portugal. The LS is a tool that helps students with the planning and monitoring of their learning experience in a course, using gamification and business intelligence techniques. Gamification was used to foster student interaction and positive competition. The LS has been initially designed to support the learning of the Data Warehouse course, which is a core subject of the Computer Science Engineering and Informatics and Management programs. This is a very demanding course in terms of study hours and practical assignments; hence, time management is critical for student success. Due to its characteristics, this course is a good case study to test the LS functionalities.

This paper describes the initial design of the LS, based on a Balanced Scorecard approach, and the prototype version of the platform, currently in use by students in the fall semester of 2016–2017. The LS is presently the research subject of two master dissertations in Computer Science Engineering, and new improved versions of the platform are scheduled to be released in the next two semesters.

2   Business Intelligence in Higher Education

Business intelligence (BI) and analytics techniques are used for data-driven decision making. BI encompasses a "broad category of applications and technologies for gathering, storing, analyzing, sharing and providing access to data to help enterprise users make better business decisions," [1]. The ultimate goal of a BI is to measure (i.e., the data related component), in order to manage, in order to enable a continuous improvement of an organization or a specific process. Hence, BI is deeply linked with performance management, requiring a positive and pro-active type of management and leadership.

An analytical mindset includes the use of data, different types of analysis (e.g., methods, approaches), and a systematic reasoning to make decisions [2]. BI and analytics are widely used in the business context, as well as in Higher Education [3, 4]. Learning analytics, is a relatively recent research area [5], focusing on the "the measurement, collection, analysis and reporting of data about learners and their contexts, for purposes of understanding and optimizing learning and the environments in which it occurs," [6].

The Balanced Scorecard (BSC) is a performance management system used to support strategic decisions. Originally developed by [7], the BSC has been successfully applied in many industries, including Higher Education (HE). Most BSC applications in HE are implemented at the institutional level, providing a performance management framework linked to the goals and strategic plans of the HE institutions. There are many examples in the literature reporting the institutional use of the BSC in academia, predominantly in the United States [8–11] and United Kingdom [12, 13], but also in many different countries [14–16]. The use of the BSC approach to strategically manage academic programs and to support the learning process is less common. Recently, [17] discusses the design of a BSC to support student success, in particular for accounting students, enabling student engagement with the educational process, as well as with the accounting profession. Other relevant examples are: [18] discussing the benefits and potential components of a BSC for an accounting program (US and Canada); [19] describing a case study, in which the BSC was applied to the Master of Business, Entrepreneurship and Technology at the University of Waterloo (Canada); and [20] describing the design of strategy maps for program performance measurement in HE. The Learning Scorecard, presented in this paper, designed according to the best practices of BI and BSC systems development, aims to measure and manage the performance and quality of the learning process. Since the LS goal is first and foremost to support students in their learning experience, it is also a valid application of learning analytics.

3 Gamification in Higher Education

Gamification is defined as the use of game design elements in non-game contexts [21]. Game elements are artifacts regularly used in game design, such as points, levels, quests or challenges, avatars, social graphs, leaderboard, badges, and rewards. Gamification, albeit being a recent trend, has been applied in several non-game contexts, such as productivity, finance, health, sustainability, and also in education [21, 22]. When using gamification the designer should keep in mind the following aspects: (1) games are to experienced voluntarily; (2) games should involve learning or problem solving; and (3) games should have some structure (i.e., game rules) but the gamer should have the freedom to explore and have fun. Barata et al. [23] present an interesting approach to gamification in education, applied to a master course in the Information systems and Computer Engineering program. In their experiment, they added a set of game (e.g., points, levels, leaderboards, challenges and badges) to the traditional course, and compared the impact of the introduced gamification on student performance and satisfaction. After a period of two consecutive years, results were very positive, with increased student performance in terms of class attendance, number of lecture slides downloads, and number of post on the course's forums [23]. This experiment was inspirational to the design of the Learning Scorecard, given that the institutional context and student profile are similar. That is, both projects are realized in Public Universities in Lisbon, Portugal, with students enrolled in similar programs (i.e., Computer Science Engineering master programs; although the LS is also being tested by undergraduate students of Information and Management).

4 The Learning Scorecard

Effective time management is pivotal for student academic success. Higher education students need to conciliate their personal and often professional responsibilities with their academic ambitions. Poor planning of tasks, in terms of effort and scheduling, often results in failing a course or achieving a lower grade than expected. In this context, the use of a strategic management tool – like the Learning Scorecard – customized by the course's faculty can provide a valuable support to students. The LS enables the formulation of strategic objectives, performance indicators and targets that support students with a baseline for the performance in the course that will yield a successful outcome. The LS also enables students to monitor their performance throughout the semester in comparison with the average performance of the entire class. This increases students' awareness of their leaning efforts, for instance, if they are falling behind the objectives defined by faculty or if they are in line with the average progression of the class. Additionally, students will be notified of incoming deadlines and their general course delay, in a proactive and motivating approach. Gamification techniques were used to design the LS, since the tool is used voluntarily by students. Gamification enables the motivation of students in terms of achieving the course goals, and provides a healthy competition environment towards the best course performance.

The LS is also a valuable tool for faculty, providing an aggregated view of students' performance and its evolution throughout the semester. The LS enables the course coordinator to identify the pain points of the course experienced by students. Namely, what are the tasks that generally demand an extra effort from students (comparing to the faculty planning) and how many students are at risk of failing the course. The analysis of student performance data can be used by the course coordinator to improve the course syllabus, with the goal of improving the teaching quality and the student learning experience.

The specification of LS included the following functional requirements:

- users need to have a profile and authentication credentials
- the LS platform needs to be integrated with the e-learning system (having access to quiz results, forum participation, downloading of materials, etc.)
- two types of accounts or views: student and course coordinator
- dashboards for monitoring individual student performance (student view)
- dashboards for monitoring class performance (course coordinator)
- automatic course schedule with deadlines and studying guidelines, customized by the course coordinator
- the LS platform should include game elements (i.e., scores, ranks, quests, leaderboard)

The design of the Learning Scorecard platform also encompasses the following non-functional requirements:

- portability across web browsers (Firefox, Google Chrome, Safari, Edge) including mobile devices
- intuitive and user-friendly interface (input data from students should be kept to a minimum)
- student identification data must be private (ethical requirements), i.e., the course coordinator will only have access to aggregated class data, even for the case of at-risk students.

The former non-functional requirement can be sensed as a miss-opportunity, since course coordinators would want to know, individually, which students are at-risk.

However, the LS was mainly designed for students to support their learning experience in the course. By introducing privacy in student identity, students are more likely to voluntarily use the Learning Scorecard and experience the benefits of this platform, without fearing any potential consequences of faculty scrutiny.

4.1   Strategic Design of the Learning Scorecard

The Learning Scorecard was designed according to the BSC methodology proposed by [24]. For the purpose of this paper, we will focus on the design decisions of a few selected steps. In the design of an organizational BSC this step usually entails the clarification of the strategy to be executed, including a strategic analysis of the organization, and the definition of mission and vision statements, as well as the organization's values. Since the LS is a thematic scorecard, the strategic analysis of the 'organization' will not be presented. The mission of the Learning Scorecard is to *"to provide Higher Education students with an analytical environment enabling the monitoring of their performance in a course, contributing to an enhanced student learning experience."* Three values encompass the design and implementation of the LS: Pursue Growth and Learning; Enjoying Participation; and Self-discipline (Make it happen). The vision statement was defined as follows: *"by the end of the academic year of 2017–2018, the Learning Scorecard application should be used by at least 50% of the enrolled students in the DSS courses at ISCTE-IUL."* The vision statement complies with the guidelines proposed by [25], in which three elements must be clearly defined: a niche (*enrolled students in the DSS courses at ISCTE-IUL*), a stretch goal (*used by at least 50% of enrolled students*) and a time frame (*by the end of academic year of 2017–2018*).

The Business Model Canvas (BMC) is a strategic tool used to describe, in an intuitive and accessible language, the business model of an organization [26], i.e., how an organization intends to create, deliver and capture value. Nine building blocks constitute the BMC: customer segments, value proposition, channels, customer relationships, revenue streams, key resources, key activities, key partnerships, and cost structure. For the purpose of the BSC design, two building blocks are of importance – customer segments and the value proposition. In order to effectively design a value proposition, matching the needs and expectations of the customer segments, [27] proposed a new canvas – the Value Proposition Canvas (VPC). In the LS we have to address the needs and expectations of both students and the course coordinator, hence we need to define two customer segments. Figures 1 and 2 present the LS value proposition canvas for students and course coordinator, respectively. Customer segments are represented on the right-end side and the value position on the left-hand side of these figures.

In the VPC model, customer segments are defined in terms of customer jobs, pains and gains. The student VPC (see Fig. 1) will be used as an example to explain the model. A customer job is related to what the customer is trying to get done; it can be a task, a problem or even a basic need (e.g., time management). The pains are the negative aspects encountered by the customer in his/her current way of dealing with the 'jobs'; including negative emotions or hurdles, undesired costs and risks (e.g., lost of

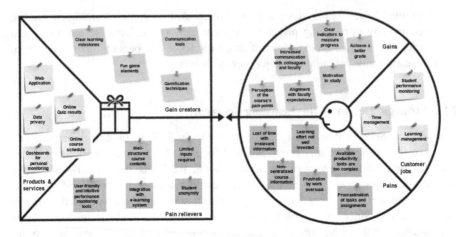

Fig. 1. Student value proposition canvas

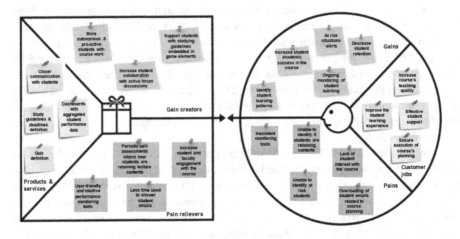

Fig. 2. Course coordinator value proposition canvas

time with irrelevant information). The gains reflect the benefits a customer expects or desires to achieve with the product or service; it can be translated into, for instance, a functional utility, positive emotions or cost savings (e.g., perception of the course's pain points). The value proposition block in the VPC is described in terms of three components: (1) products and services (a list of products and services offered and their link to the customers' jobs); (2) pain relievers (to eliminate or reduce the customers' pains); and (3) gain creators (describing the positive outcomes and benefits that products and services deliver to customers). An example of these components are, respectively, (1) online course schedule, (2) well-structured course contents, and (3) clear learning milestones.

The observation and design of the customer segment profile comes first. Then follows the design of the value proposition, addressing the most relevant and critical

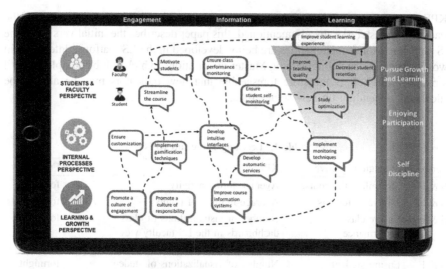

Fig. 3. The Learning Scorecard strategy map

jobs, pains and gains of the target customers. In this way, the design process of the product/service is enclosed by the real needs of customers and there is a concrete mapping with expected benefits. This process is also useful to determine the differentiator factors of the product/service, which will be useful for the definition of performance indicators in the BSC.

The strategy map is a design tool [8] that helps to tell the story of the strategy, highlighting in a creative way the dependencies, called cause-and-effect relationships, between the strategic objectives. The map, displayed in Fig. 3, has three perspectives: Students and Faculty (S & F), Internal Processes (IP), and Learning and Growth (L & G). The financial perspective, the fourth standard perspective in a Balanced Scorecard is not relevant for the Learning Scorecard, which is focused solely on the student learning experience. Three strategic themes were also defined to frame the definition of strategic objectives: Engagement, Information and Learning. These themes are the main drivers to achieve the vision. That is, if we aim to have the LS platform being used by at least 50% of all enrolled students, then it is key to foster student engagement, provide updated information of performance monitoring, and help students to optimize their learning experience in the course.

The strategy map should be read bottom-up, following the cause-and-effect relationships between objectives. As can be seen in Fig. 3, the final strategic objective (the ultimate effect) is to improve the student learning experience, which is the central goal of the Learning Scorecard. Values are included in the strategy map, at the right end side of the "mobile phone", a metaphor for the portability of the LS platform, since the development of a mobile application is contemplated in the near future.

A crucial part of the balanced scorecard is the definition of performance indicators to measure the achievement of the intended strategic objectives. In this project, a subset of these performance indicators will also be used to populate the student and course coordinator dashboards provided by the LS platform. The key performance indicators

(KPI) of the LS, presented in Table 1, are either measured biweekly or at the end of the academic semester. As already mentioned, this paper describes the initial version of the LS platform. More functionalities are being developed in the LS platform, integrated in two Computer Science Engineering Master dissertations, which will be completed until September 2017. It is therefore foreseeable that new indicators may be able to be calculated, depending on new source data.

Table 1. Learning Scorecard KPIs

Persp.	Strategic objective	KPI	Frequency
S & F	Streamline the course	Average forum activity	fortnight
S & F	Motivate students	Average number of points	fortnight
S & F	Ensure class performance monitoring	Number of visualizations of class dashboards in the LS faculty view	fortnight
S & F	Ensure student self-monitoring	Number of visualizations of student dashboards in the LS student view	fortnight
S & F	Study optimization	% completed quests within course milestones	fortnight
S & F	Improve teaching quality	Average final grade	semester
S & F	Decrease student retention	Course retention rate	semester
S & F	Improve student learning experience	Student satisfaction index	semester
IP	Ensure customization	% of used LS input options	semester
IP	Implement gamification techniques	Number of LS game elements	semester
IP	Develop intuitive interfaces	Student usability assessment index	semester
IP	Develop automatic services	% quests without manual user input	semester
IP	Implement monitoring techniques	Number of LS data visualizations	semester
L & G	Promote a culture of engagement	% LS active students	fortnight
L & G	Promote a culture of responsibility	Average quest delay	fortnight
L & G	Improve course information systems	Number of new implemented LS functionalities	semester

5 LS Prototype

The LS platform was developed using Node.js. The front-end was developed using HTML and CSS. Javascript, specifically Express.js, was used for the back-end implementation. Several modules were used: Bootstrap, for platform design, Chart.js, for the implementation of the charts in the LS dashboards, Passport and Crypto, for secure authentication of students in the LS. The LS pilot also includes a MySQL database that stores data from the e-learning system and custom data provided by the course coordinator (input format .csv).

110 students are currently testing the pilot implementation of the Learning Scorecard in the Data Warehouse course; in the 2016–2017 fall semester this course is offered to four different programs. In the LS, students are divided into classes according to their program. Students begin with zero points, and are thus encouraged to learn to earn points, and increase their game level. A summary of the leaderboard is always visible (at the left down corner of Fig. 4), presenting not only the top-5 gamers (ordered by points) but also the ranking of classes in terms of the percentage of active students using the LS. In the ranking, each class is identified by the respective program acronym (in Portuguese): MEI, IGE, IGE-PL, and METI. Figure 4 presents the initial page of the LS for a student (i.e., a "gamer"). In this page the student can visualize his/her performance (in points) and receive alerts about incoming quests deadlines. Since the first LS experiment in ongoing, Figs. 4, 5, 6, 7, 8 and 9 display real data as of November 18, 2016 (week 9 of the semester, which comprises 12 weeks), kindly sent by students to the course coordinator for the purpose of this paper. Figure 5 displays the detail of the leaderboard, with the identification of top gamers.

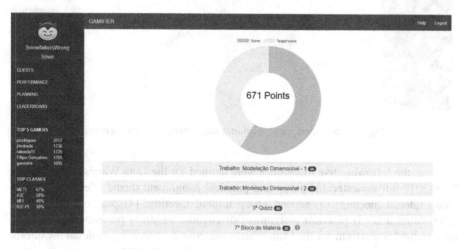

Fig. 4. LS student view: quests and points

The planning functionality is developed based on the course syllabus (currently still in Portuguese). Each semester, the course coordinator needs to customize the set of quests as well as course's milestones, and their respective deadlines or due dates. For

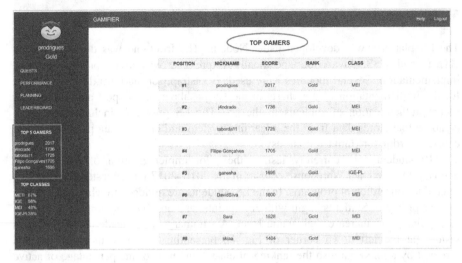

Fig. 5. LS student view: Leaderboard

Fig. 6. LS student view: planning functionality (in Portuguese)

instance, the following five milestones were defined for the Data Warehouse course in 2016-2017 fall semester: (1) group and practical assignment theme selection; (2) first group tutorial meeting; (3) second group tutorial meeting; (4) practical assignment submission; and (5) practical assignment discussion. Figure 6 presents the planning page in the LS platform for the Data Warehouse course, describing in detail the quests for weeks 9 and 10. By clicking on each quest in this list, the student can visualize a pop-up interface with a detailed description of the quest, aligned with the information of the syllabus, and the number of points that can be awarded. Quests can be mandatory or optional. It is also possible to customize how many points students loose if they fail to comply with the quest deadline. For mandatory quests, the deadline must be met, otherwise no points are awarded. Quests are related to reading of lectures' slides, solving exercises, completing the milestones of the practical assignment, class attendance, solving quizzes, participating in the course's forum, etc.

The performance functionality in the student view includes three standard visualizations: progress analysis, percentage chart and radar chart, represented in Figs. 7, 8 and 9, respectively. These figures display data for the top gamer at the time (*prodrigues*) and his performance until week 9 of the semester.

The progress analysis view (Fig. 7) is composed by a set of charts reporting on points earned with quizzes, class attendance, practical assignment and lecture readings, as well as forum participation. As the semester unfolds, these charts are automatically adjusted to report on the performance of each week. As can be observed in Fig. 7, in

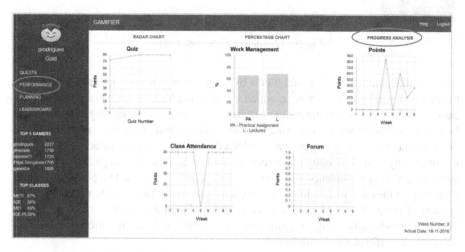

Fig. 7. LS student view: performance visualization (progress analysis)

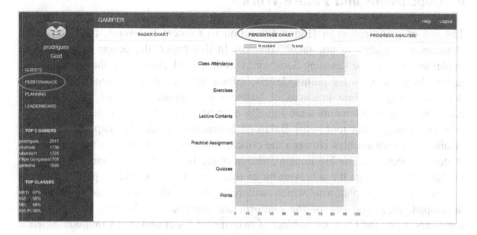

Fig. 8. LS student view: percentage chart

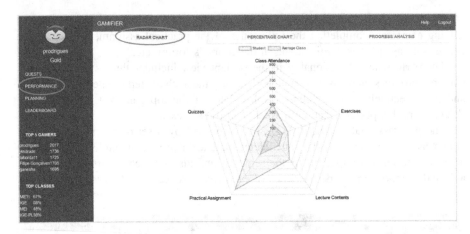

Fig. 9. LS student view: radar chart

the chart that summarizes the total points by week, there is no performance reported on the first four weeks of the semester, which was due to a hacking problem. The Learning Scorecard database was compromised during the cyber attack, leading to a complete loss of activities. Students were asked to redo all the quests until the day of the attack and the security of the platform was reinforced. Another relevant aspect is the lack of forum activity. This is one topic that requires further development, especially on how to motivate and trigger students' discussions on the course's forum. The percentage chart (Fig. 8) provides a visualization of the student's current achievements versus the total amount of points that he/she could have earned so far in the different types of quests. Finally, the radar chart (Fig. 9) compares the student's individual performance versus the average performance in his/her class.

6 Conclusions and Future Work

The use of gamification in Higher Education is a recent technique, in which game elements are applied to non-game contexts. In this paper, the design and prototype implementation of the Learning Scorecard was described. Apart from the integration of game elements, such as points, levels, quests and leaderboard, the LS was also designed using the best practices of business intelligence. A set of functional and non-functional requirements were initially defined, which led to the definition of the strategic management tool – the Balanced Scorecard. A strategy map was designed detailing the relationships between the critical components of the LS strategy, i.e., the strategic objectives. The full implementation of the BSC, with dashboards to visualize the KPIs that measure the achievement of strategic objectives, will be part of the next version of the LS platform, which will have the course coordinator's view fully developed. New versions of the platform are planned for the next two academic semesters, since the LS is the subject of two master dissertations in Computer Science Engineering that are due September 2017.

Currently, the LS platform is being used by 110 students in a pilot study focused on the usage of the LS student view in one course. The platform is already integrated with the e-learning platform Blackboard used at the university. The selected course – dedicated to the development of Data Warehouse systems – is a very demanding course in terms of study hours and the complexity of the practical assignment. Initial data reports on a percentage of active students in each class using the LS platform in a range between 38% and 67%. A deeper analysis of the two largest classes in terms of number of students (accounting for approximately 85% of the total number of enrolled students) reveals that: (1) one class (IGE), from the Information and Management bachelor program, has more active students (58%), but a lower point average; (2) another class (MEI), from the Computer Science Engineering master program, has comparatively a lower number of active students (48%), but with a higher number of points (in average). Therefore, we can conclude from the initial data that MEI students have so far a greater commitment to the usage of the LS platform.

Future work entails the identification of study patterns linked to student success, using data mining techniques. The integration with Blackboard will also be further explored, particularly in terms of collaborative learning aspects already present in the tool, such as the use of the forum. The gamification part will also be extensively developed, specifically in terms of visual effects. The LS aims to be a fun tool, that really makes a difference in the way students learn and collaborate with each other. It is also planned the design and implementation of student satisfaction questionnaires, to assess student engagement, motivation, and satisfaction with the course and the LS platform. Student motivation is vital for the success of this platform. Further research also entails the design of motivation and reward mechanisms integrated with game dynamics to increase the number of active students as well as the number of completed quests and points earned.

Acknowledgments. This work has been presented at the 2nd Workshop on Educational Knowledge Management [28] and has won the best paper award. This workshop was co-located with the 20th International Conference on Knowledge Engineering and Knowledge Management (EKAW 2016), in Bologna, Italy, November 19-23, 2016.

References

1. Burton, B., Geishecker, L., Schlegel, K., Hostmann, B., Austin, T., Herschel, G., Soejarto, A., Rayner, N.: Business Intelligence focus shifts from Tactical to Strategic. Gartner Research Note G00139352 (2006)
2. Davenport, T., Harris, J., Morison, R.: Analytics at Work – Smarter Decisions, Better Results. Harvard Business Press, Boston (2010)
3. Rajni, J., Malaya, D.B.: Predictive analytics in a higher education context. IT Prof. **17**(4), 24–33 (2015). IEEE Computer Society Publishing
4. van Barneveld, A., Arnold, K.E., Campbell, J.P.: Analytics in Higher Education: Establishing a Common Language. Educause (2012)
5. Clow, D.: An overview of learning analytics. Teach. High. Educ. **18**(6), 683–695 (2013)
6. Long, P., Siemens, G.: Penetrating the fog: analytics in learning and education. Educause Rev. **46**(5), 31–40 (2011)

7. Kaplan, R., Norton, D.: The Balanced Scorecard – Translating Strategy into Action. Harvard Business School Press (1996)

8. Kaplan, R., Norton, D.: Having trouble with your strategy? then map it. Harvard Bus. Rev. **78**, 167–176 (2000)

9. O'Neil Jr., H., Bensimon, E., Diamond, M., Moore, M.: Designing and implementing an academic scorecard. Change **31**, 33–40 (1999)

10. Karathanos, D., Karathanos, P.: Applying the balanced scorecard to education. J. Educ. Bus. **80**, 222–230 (2005)

11. Beard, D.: Successful Applications of the Balanced Scorecard in Higher Education. J. Educ. Bus. **84**, 275–282 (2009)

12. Thomas, H.: Business school strategy and the metrics for success. J. Manage. Dev. **26**, 33–42 (2007)

13. Taylor, J., Baines, C.: Performance Management in UK universities: implementing the Balanced Scorecard. J. High. Educ. Policy Manage. **34**(2), 111–124 (2012)

14. Boned, J.L., Bagur, L.: Management Information Systems: The Balanced Scorecard in Spanish Public Universities (2006). Available at SSRN: http://ssrn.com/abstract=1002517

15. Yu, M.L., Hamid, S., Ijab, M.T., Soo, H.P.: The e-balanced scorecard (e-BSC) for measuring academic staff performance excellence. High. Educ. **57**, 813–828 (2009)

16. Pietrzak, M., Paliszkiewicz, J., Klepacki, B.: The application of the balanced scorecard (BSC) in the higher education setting of a Polish university. Online J. Appl. Knowl. Manage. **3**(1), 151–164 (2015)

17. Fredin, A., Fuchsteiner, P., Portz, K.: Working toward more engaged and successful accounting students: a Balanced Scorecard approach. Am. J. Bus. Educ. **8**(1), 49–62 (2015)

18. Chang, O., Chow, C.: The balanced scorecard: a potential tool for supporting change and continuous improvement in accounting education. Issues Account. Educ. **14**, 395–412 (1999)

19. Scholey, C., Armitage, H.: Hands-on scorecarding in the higher education sector. Planning High. Educ. **35**, 31–41 (2006)

20. Cardoso, E., Viaene, S., Costa, C.S.: Designing strategy maps for programme performance measurement in higher education. In: 15th International Conference of European University Information Systems (EUNIS 2009), Spain (2009)

21. Deterding, S., Dixon, D., Khaled, R., Nacke, L.: From game design elements to gamefulness: Defining "Gamification". In: 15th International Academic MindTrek Conference, Finland (2011)

22. Dicheva, D., Dichev, C., Agre, G., Angelova, G.: Gamification in Education: a systematic mapping study. Educ. Technol. Soc. **18**(3), 75–88 (2015)

23. Barata, G., Gama, S., Jorge, J., Gonçalves, D.: Engaging engineering students with gamification. In: 5th International Conference on Games and Virtual Worlds for Serious Applications (vs-games) (2013)

24. Cardoso, E.: Performance and Quality Management of Higher Education Programmes. Ph.D. thesis, University Institute of Lisbon (ISCTE-IUL) (2011)

25. Kaplan, R., Norton, D.: The Execution Premium – Linking Strategy to Operations for Competitive Advantage. Harvard Business School Press, Boston (2008)

26. Osterwalder, A., Pigner, Y.: Business Model Generation – a Handbook for Visionaries, Game Changers, and Challengers. Wiley (2010)

27. Osterwalder, A., Pigner, Y., Smith, A., Bernarda, G., Papadakos, P.: Value Proposition Design. Wiley (2014)

28. Cardoso, E., Santos, D., Costa, D., Caçador, F., Antunes, A., Parada, R.: Learning Scorecard: monitor and foster student learning through gamification. In: 2nd International Workshop on Educational Knowledge Management, Italy. CEUR Workshop Proceedings, vol. 1780, pp. 39–50 (2016)

Towards an Architecture for Universities Management

Assisting Students in the Choice of Their Specialization

Inaya Lahoud[1]([⊠]) and Fatma Chamekh[2]

[1] Department of Computer Science,
University of Galatasaray, Istanbul, Turkey
clahoud@gsu.edu.tr
[2] University of Lyon, Lyon, France
Fatma.chamekh@univ-lyon3.fr

Abstract. The heterogeneity and the high volume of data on the Web are the main features that make it a promoter field of knowledge engineering for researcher. However, the user is getting lost towards the diversity of information on the Web. In this paper, we propose an approach, to assist user, based on Semantic Web technologies. Our scenario is focused on the field of education and in particular higher education. This choice is motivated by the diversity of information sources where the student is dispersed.

Keywords: Ontologies · Higher education institution · Universities management system

1 Introduction

Nowadays, the web offers advanced interactions between users and data providers. Since the open data initiative, the data on the Web has increased. Many institutions have published their data in heterogeneous and distributed way. Universities and educational institutions follow this way by providing a data concerning their training offers. Users retrieve and review educational training. The main idea is to share their experiences and get the best feedback about universities.

Our Objective is to match the user's needs and trainings offered by institutions. The management of data from diverse resources is complex and tedious process, relying mainly on human-based and error-prone tasks. Globally, there is a lackof practical approach for converting and linking multi-origin data pieces into one coherent smart data set. More specifically, the following scientific locks make the transformation of data on a smart data a difficult task. For this end we present the main challenges behind our research work:

- Data usually comes in variable quality unorganized or not described, and not linked to other sources on the Web. There is a need to homogenize vocabularies by providing machine-readable explicit description of data semantics, and linking data sets to each other and to ontologies.

P. Ciancarini et al. (Eds.): EKAW 2016 Satellite Events, LNAI 10180, pp. 69–81, 2017.
DOI: 10.1007/978-3-319-58694-6_6

- The combination of heterogeneous data from different sources generates some issues. Those one could be classified on three levels: the syntactic level related to data format and syntactic, the semantic level when different knowledge presentations are used and the structural level due to the different data organizations.

We focused in this paper on assisting users in their information research by offering them a universities management system. This system seeks the most appropriate training for a student according to his criteria (university ranking, domain of training, geography of university and unemployment percent). In addition, this system offers to him trainings where he can benefit from social or excellence scholarships, grants from the state to encourage certain areas, or simply where the unemployment rate is lowest and therefore he has more chances to acquire employment at the end of his studies. Our system uses the Open Data of higher education and unemployment from the data.gouv.fr website.

The remainder of this paper is organized as follows: Sect. 2 states our problem, Sect. 3 presents some existing approach that use semantic web technologies for education. Section 4 depicts our framework. Future works and conclusion are presented in Sect. 5.

2 Problem Statement

We conducted our study in the field of education and more specifically on higher education institutions. Indeed, the number of higher education institutions is increasing every year. If we count the number of these institutions in a single country such as France (more than 3500[1]) and the United States (more than 4500 [1]), and the number of training courses that offer these institutions, we can understand why students are always lost in the choice of their studies.

The final year of high school is the year when students accrue the stress of the passage of the high school diploma, the daily difficulties and the choice of their future orientation.

Most students, who have managed to continue their studies, think that they were insufficiently or badly informed and advised and therefore they chose the training in which they can succeed. Not having a clear and precise idea of the job they wish to exercise, nor a long-term professional objective, these students eventually chose more often the curriculum that leaves most open doors. Nevertheless, the difficulties are not especially evacuated for those who are still looking for jobs, then they wonder about their chances to integrate quickly the job market and are asking questions about the appropriateness of their courses' choice [2]. Indeed, according to Gordon [3], an estimated 20 to 50% of students enter college as "undecided" and 75% change their major at least once before graduation. Furthermore, the Office of Institutional Research in Western Kentucky University published a report in 2012 entitled "Does Changing Majors Really Affect the Time to Graduate? The Impact of Changing Majors on Student Retention, Graduation, and Time to Graduate" [4]. They found that one-third of students in their university chose an initial "undeclared" major. These students will

---

[1] education.gouv.fr.

have no problem if they declared a specific major by the end of the second year. Delaying the choice of a major after the second year resulted in rapid deterioration of success rates.

In front of this situation, we need to propose to students a system able managing a huge and heterogeneous amount of data and take into account their domain of interests and preferences. Our general challenge is summarized as follows: given the data provided by educational institutions, government and companies, whose understand-ability remains difficult, semantic web technologies could provide an efficient solution.

3   Related Work

The exploitation and the search for information on the Web become increasingly complicated task. How to find information that we seek quickly given the diversity and the quantity of data on the Web? Where to search this information since there is no global database that stores all data of the whole world as wished Tim Berners Lee? What is the relevance of the information we found on the Web? Do they fully correspond to our research?

There are many researchers working on these issues and try to find solutions. As we explained in the introduction, we focused our study on the education and specifically on higher education institutions. Our goal is to consolidate information of higher education in one system where students would find what training is the most appropriate for them basing on university ranking, percentage of employment in this field, geography, and available scholarships.

Semantic web technologies are getting increasingly used in various contexts, higher education is no exception. Different approaches are designed to spread the services of the universities which are distributed across several departments to serve their substantial student base. In their survey, Dietze et al. [5] highlight the growing use of linked data by various universities. Many platforms are made available for direct consumption and reuse. This includes for example the OU's linked open data platform (http://data.open.ac.uk), the University of Muenster (http://data.uni-muenster.de), the University of Southampton (http://data.southampton.ac.uk) among others. The data available through those platforms include: Courses information podcasts, Library catalogue, research publications, OpenLearn, reading experience database, the open arts archive events, information about university staff and buildings located across the UK. Besides that, the data of the platforms cited above are connected to the linked open data cloud. Dedicated graphs include links to the Dbpedia entities, Geonames entities and BBC entities. The external entities have the topics of media objects, web pages, courses.

The last few years, we have seen several websites style MOOCs such as Coursera[2] FUN[3]. These websites allow users a free access to online courses. However, the aim of

---

our system is to search trainings offered by universities and not online courses as the MOOC case.

[6–8] have proposed methodologies and frameworks to transform the existing data sources into linked data. The main idea is to turn the available organizational data in a linked data cloud, using pre-programmed transformation pattern. To follow the success of social and knowledge graphs functionalities provided by facebook[4] and google[5], Health et al. [9] proposed to create an education graph by processing courses information and learning material from various universities in UK. They rely on bibliographical data of material repositories to identify links to course resource. In this direction, the LinkedUpDataCatalog[6] or related community initiatives[7] are initial efforts to collect and catalog dataset have been made by universities or other institutions.

Zablith [10] created a semantic data layer to conceptually connect courses taught in a higher education program. The aim is to interlink courses within the same institution at the level of concepts covered in course topics. For this reason, the author proposed a courses information data model.

The limitations of approaches explained above are mainly related to the following aspects. First, the data provided by those platforms are mainly limited to the educational information such as courses program, publication. That means student could find the courses topics and courses material or web pages. Also, those dataset are related to external dataset such as Dbpedia or geonames but to enrich the educational data. Second, the first efforts to connect various datasets are proposed by [9] but it is limited to universities of UK.

In the other side we didn't find researches who occur the subject of trainings' classification to obtain scholarships except Rad [11] who presented in his paper a study in Iran on the classification of university courses using data mining techniques. These courses were classified according to different criteria to determine which training is the most promising in the future. The Iranian state uses the result of this classification then to invest in these trainings and provide scholarships to encourage students to follow them.

In our context, the aim is to assist the student in the selection of training. This selection should be made following various criteria such as the university rank, scholarships existence, and the unemployment average. To our knowledge, there is no existing approach that covers these requirements. To achieve our objective, we have to connect the educational and government datasets. We present in the next section our approach of universities management system in order to assist students in their choice.

---

[4] http://newsroom.fb.com/News/562/Introducing-Graph-Search-Beta.

[5] http://www.google.com/insidesearch/features/search/knowledge.html.

[6] http://data.linkededucation.org/linkedup/catalog.

[7] http://www.w3.org/community/opened.

4 Global Approach

The idea of our work is to deal with academic and government data and make them more accessible to users. Our methodology consists of four main steps: knowledge definition, linked open data, knowledge extraction and classification, and knowledge representation (Fig. 1). We use the word "knowledge" in our system because we enrich the data with semantic information.

The first step in our system is to backup data in ontological format. This allows us to represent them formally and semantically. The data used in our study are open data from the Website data.gouv.fr. Once the data are well represented, we take the two ontologies, in the second stage, and we try to link them with the online data as dbpedia, and others. All OWL files are saved in a knowledge base.

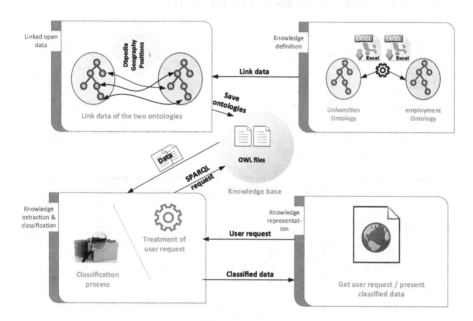

Fig. 1. Our global approach

The first three steps are in the systems level only while the fourth is in interaction with the human being. Indeed, the human being formulates his query in the search engine and sends it to the system. This request is received by the system in the third step of our methodology (knowledge extraction & classification). It treats the user's query and transforms it into SPARQL query in order to apply it on our OWL related files. The resulting data, from this extraction, will be sorted and sent back to the fourth step to present them to the user.

Currently the first two steps are performed only once. We do not manage the evolution of ontology over time. Whereas the last two steps are triggered for each request sent by user.

4.1 Knowledge Definition

As we mentioned before, we use the data available on the government website. These data can be extracted from this website only in Excel format (Fig. 2). To represent the data in a more formal and semantic way, we have chosen to transform them into OWL language. This language allows to describe ontologies, that is to say, it defines terminology to describe specific areas. Ontologies have shown in recent years their ability to model a range of knowledge in a given field. The transformation of our data in OWL was done with an open source software RDBToOnto [12].

	A	B	C	D
1	uai	uo_lib	sigle	type_d_etablissement
2	0311382J	Université Toulouse 1 - Capitole		Université
3	0530311T	école nationale supéri ENS Chimie Lill		écoles nationales supéri
4	9840343G	Université de la Polyné UPF		Université
5	0350095N	école des hautes étud EHESP		Grand établissement
6	0630187D	école centrale de Lyon EC Lyon		Instituts et écoles extérie
7	0133774G	école centrale de Marseille		Instituts et écoles extérie
8	0912274C	UniverSud Paris		Communauté d'universit
9	0753486G	école pratique des hau EPHE		Grand établissement
10	0341087X	Université Montpellier 1		Université
11	13400048	Casa de VelÃ¡zquez de Madrid		école française Ã l'étran
12	0352440M	école normale supérie ENS Rennes		école normale supérieu
13	0861286H	Centre national d'ense CNED		Autre établissement
14		Collegium Alle-de-France		Communauté d'universit
15	0440100V	Ecole centrale de Nantes		Instituts et écoles extérie
16		Consortium national po AGREENIUM		PÃ´le de recherche et d'i

	A	B	C	D	E	F	G
1	Diplôme et spécialité de formation		Part des fem	Taux de chômage	Part d'emplois à temps	Part des cadres et profi	Salaire médian en euros 2009
2	Non diplômés, certificat d'études primaires, brevet des collège		37	31	19	11	1 130
3	CAP ou BEP et équivalent en Agriculture, pêche, forêt, espaces		20	13	11	6	1 220
4	CAP ou BEP et équivalent en Agro-alimentaire, cuisine		17	14	9	5	1 240
5	CAP ou BEP et équivalent en Génie civil, construction, bois		2	15	2	6	1 280
6	CAP ou BEP et équivalent en Textile, habillement, cuir		58	27	19	3	1 080
7	CAP ou BEP et équivalent en Mécanique		2	14	4	9	1 290
8	CAP ou BEP et équivalent en Électricité, électronique		3	15	6	14	1 300
9	CAP ou BEP et équivalent en Commerce, vente		70	24	30	8	1 060
10	CAP ou BEP et équivalent en Finances, comptabilité, gestion		55	23	22	9	1 140
11	CAP ou BEP et équivalent en Secrétariat, communication		85	25	28	14	1 100
12	CAP ou BEP et équivalent en Accueil, hôtellerie, tourisme		64	23	25	12	1 100
13	CAP ou BEP et équivalent en Coiffure, esthétique		92	20	24	3	1 040
14	Dip. paramédical et social de niveau CAP-BEP (aides-soignante		93	6	15	2	1 350
15	Bac professionnel et équivalent en Agriculture, pêche, forêt, e		22	6	9	10	1 190
16	Bac professionnel et équivalent en Agro-alimentaire, cuisine		20	8	4	15	1 320
17	Bac professionnel et équivalent en Génie civil, construction, b		6	5	2	16	1 370
18	Bac professionnel et équivalent en Mécanique		2	7	2	24	1 400
19	Bac professionnel et équivalent en Électricité, électronique		2	9	2	32	1 410

Fig. 2. Two excerpts of data from data.gouv.fr website

Before the transformation of data into ontologies we clean them to optimize their quality whether that of higher education institutions or employment. Our data cleaning process will affect incomplete, noisy and inconsistent data. For instance, we can see in the first excerpt of Fig. 2 that we have two missing UAI for two universities. This number is the ID of the university so we had to search this information and add it to the document. We formalized the phone, and SIRET numbers. We split the coordinate information into longitude and latitude one in order to link them to DBpedia. Once the transformation process finish, a verification process has been performed manually by a human expert in order to verify the quality of RDBToOnto results.

Our higher education ontology contains now information about the institutions however it doesn't provide us any information about trainings they offered. We tried firstly to develop a program to search this information directly from the website of each university and include it to our ontology. We found that it is impossible to search the trainings information directly from their website because there is no common standards. Indeed, each university has its own standards, template, and vocabularies. Thus, if we search the word "formation" in the URL, it is impossible to get the needed information among the thousands of obtained results. In our case, we found that the better solution was to use the catalog of trainings from "CampusFrance" website. It provide us all trainings in bachelor and master levels. Therefore, we developed a JAVA program to extract this information and include it in our higher education ontology.

4.2   Linked Open Data

This step consists of taking the OWL files from the knowledge base and link them with open data available online such as the geographical data from INSEE. Indeed, the two OWL files on which we work contain geographical data. Therefore, we have linked these data with the geographical data of INSEE as in the following example.

Figure 3 shows the geographic data (municipality, department, and region) of a university and the rate of employment/unemployment by region and how we related them to the open data on the Web.

Then we have linked other data such as areas or trainings. As the employment rate is also classified by domain (agriculture, commerce, computer, construction ...), and institutions offer trainings (IT, networks, mechanical, civil Engineering, political science, medicine ...) that are applied in specific areas, we can well link these two fields (Fig. 4). However, these data are not available in open data until now so we created our own links to connect them.TrainingLevel corresponds to Master, Bachelor, and Engineer degree.

4.3   Knowledge Extraction and Classification

This step is divided into two distinct parts: the extraction and classification.

In the first part, the system retrieves the request submitted by the user as shown in (Fig. 6) and reformulates it in SPARQL query language (Fig. 5) to execute it on OWL files. The result of this query will then be classified by the most relevant to user (Fig. 7). Indeed, given a user query, traditional search engines output a list of results which are ranked according to their relevance to the query. However, the ranking is independent of document topic. Therefore, the results of different topics are not grouped together within the result output from a search engine. This can be problematic, as the user must scroll though many irrelevant results until his desired

```
<Universite rdf:ID="Autre_établissement_12">
      <owl:sameAs
rdf:resource="http://fr.dbpedia.org/page/Centre_national_d'enseignement_%C3%A0_distance"/
>
      <statut_juridique_long>établissement          public          à          caractère
administratif</statut_juridique_long>
      <uo_lib>Centre national d&apos;enseignement à  distance</uo_lib>
      <identifiant_freebase>http://www.freebase.com/m/0d98ld</identifiant_freebase>
      <code_postal_uai>86960</code_postal_uai>
                  <aCommeFormation            rdf:resource="#Licence_professionnelle_-
_Droit,_économie,_gestion_-_hôtellerie_et_tourisme_-
_spécialité_chef_de_projet_et_créateur_d'entreprises_touristiques" />
                  <aCommeFormation            rdf:resource="#Licence_professionnelle_-
_Sciences,_technologies,_santé_-_mécanique_-
_spécialité_coordinateur_technique_des_méthodes_d'industrialisation" />
                  <aCommeFormation
rdf:resource="#Diplôme_d'ingénieur_du_Conservatoire_national_des_arts_et_métiers_spécialit
é_aéronautique_et_spatial_en_convention_avec_l'ISAE-
ENSMA,_en_partenariat_avec_AEROTEAM" />
      ........
      ........
      ........
      <libelle_mission_chef_de_file>Enseignement scolaire</libelle_mission_chef_de_file>
      <uai>0861288H</uai>
      <lieu_dit_uai>ASTERAMA 2 AV DU TELEPOR</lieu_dit_uai>
      <localite_acheminement_uai>CHASSENEUIL CEDEX</localite_acheminement_uai>
      <uucr_id>UU86601</uucr_id>
   element_wikidata>http://www.wikidata.org/entity/Q2350714</element_wikidata>
      <boite_postale_uai>300</boite_postale_uai>
      <url>http://www.cned.fr/</url>

<identifiant_programme_lolf_chef_de_file>214</identifiant_programme_lolf_chef_de_file>
      <statut_juridique_court>EPA</statut_juridique_court>
      <statut_operateur_lolf>Opérateurs LOLF Hors MIRES 2014</statut_operateur_lolf>
      <uucr_nom>Poitiers</uucr_nom>
      <geo:Departement rdf:about="DEP_86">
                  <geo:code_departement>86</geo:code_departement>
                  <geo:nom xml:lang="fr">Vienne</geo:nom>
      </geo:Departement>
      <coordonnees>
                  <gs:long>46.6512</gs:long>
                  <gs:lat>0.372263</gs:lat>
      </coordonnees>
      <aca_id>A13</aca_id>
      <flux_rss>http://www.cned.fr/rss/communiques-de-presse.xml</flux_rss>
      <numero_telephone_uai>0549493400</numero_telephone_uai>
   </Universite>
```

Fig. 3. Extract from higher education ontology

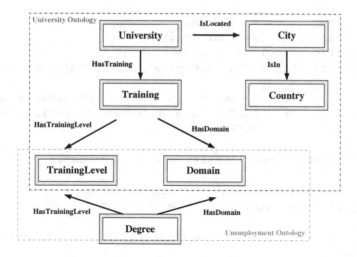

Fig. 4. Linking university and unemployment ontologies

information need is found. This might arise when the user is a novice or has superficial knowledge about the domain of interest, but more typically, it is due to the query being short and ambiguous. Therefore, to avoid this problem we will present in the future a classification method for results search to satisfy the request of user.

4.4 Knowledge Representation

This step allows us to offer a website to the user where he can submit his request for a search and receive the corresponding information. The system then retrieves the query when the user submits it and sends it to the step "Knowledge Extraction & classification" to treat it. The system receives by return the query result already classified and displayed it to the user by order of relevance.

In this example, the student search for a training in chemistry, commerce or law domain in four regions in France.

Figure 7 shows the result of his search. It is classified by domain. In each one, student can find the name of training and the university that offers it. The system shows also the percent of unemployment by domain and by training level depending on the information we extracted from the data.gouv.fr website.

We didn't implement yet the classification system to classify results in each domain by the most relevant to the student.

```
String sparqlQuery = "PREFIX
ontologie1:<http://www.universities-
project.fr/ontologies/universites#>\n" +
"PREFIX rdf:<http://www.w3.org/1999/02/22-rdf-syntax-
ns#>\n" +
"PREFIX rdfs:<http://www.w3.org/2000/01/rdf-schema#>\n"+
                "SELECT ?nom_univ ?nom_formations \n" +
                "WHERE {\n" ;

  if(!domaine1.equals("Choisissez_un_domaine"))
            {
            sparqlQuery += " ?formations
ontologie1:hasDomaine ontologie1:"+ domaine1 +"  .\n" ;
            }
            else // Si l'utilisateur demande tous les
domaines
            {
                    lblNomdomaine.setText("Tous les
domaines");
            panel_res_3.setBounds(0, 0, 785, 460);
            }
...
...
{sparqlQuery += "?formations
ontologie1:hasNiveauFormation ontologie1:"+ ite +" .\n";}
...
...
sparqlQuery += "?universites ontologie1:aCommeFormation
?formations .\n";
...
...
sparqlQuery += "{ ?universites ontologie1:new_reg_nom
\""+ region + "\" } UNION ";
...
...
}
```

Fig. 5. Excerpt of SPARQL query

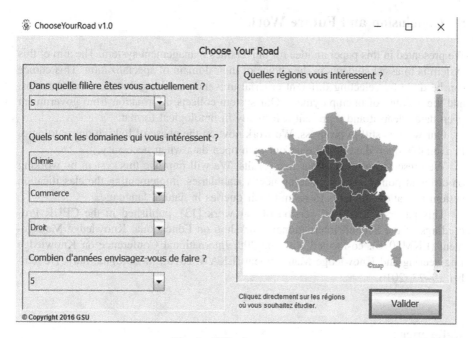

Fig. 6. GUI of our system

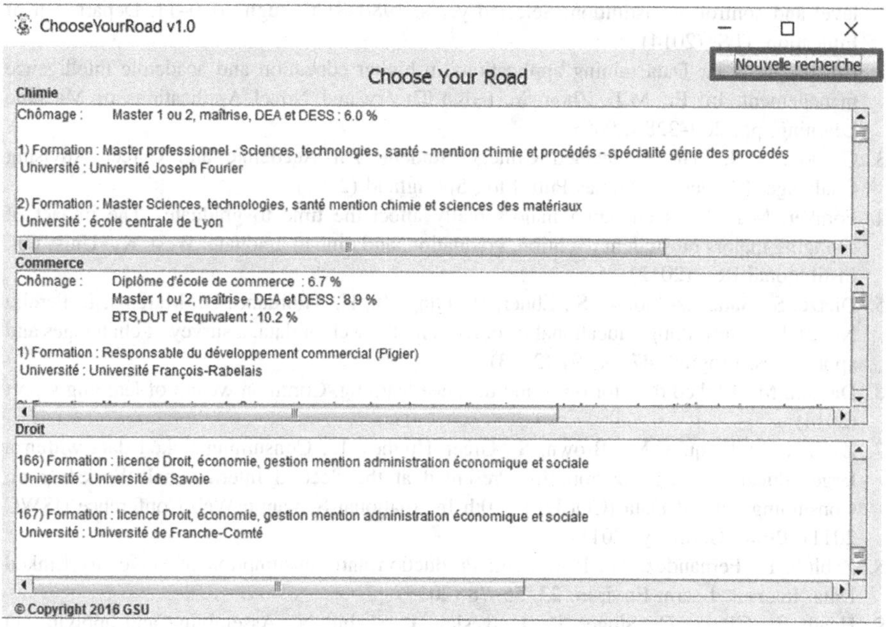

Fig. 7. Results of research

5 Conclusion and Future Work

We presented in this paper an idea for universities management system. The aim of this system is to assist students in the choice of their domain of specialization. This choice can be done by selecting different criteria: university ranking, location, scholarships, and the average of unemployment. Our system collects information from government open data, clean it and represent it formally in ontological format.

Our work is still in progress. We work now on the phase of linking our ontologies to available open data or creating our own open data when necessary.

We presented in Sect. 4 our first results. We will improve this system by working on different points such as creating new vocabularies, implementing the classification system and allow students to submit their queries in natural language.

This paper is a revised version of our work [13], published in the CEUR Proceedings of the Second International Workshop on Educational Knowledge Management (EKM2016), co-located with the 20th International Conference on Knowledge Engineering and Knowledge Management (EKAW 2016), in Bologna, Italy, November 19–23, 2016.

References

1. National Center for Education Statistics: Table 5 Number of educational institutions, by level and control of institution: Selected years, 1980–81 through 2010–11. Department of Education, U.S. (2014)
2. Bresfelean, V.P.: Data mining applications in higher education and academic intelligence management. In: Er, M.J., Zhou, Y. (eds.) Theory and Novel Applications of Machine Learning, pp. 209–228 (2009)
3. Gordon, V.N.: The Undecided College Student: An Academic and Career Advising Challenge. Charles C Thomas Pub. Ltd., Springfield (2007)
4. Foraker, M.J.: Does changing majors really affect the time to graduate? The impact of changing majors on student retention, graduation, and time to graduate. West. Ky. Univ. Off. Institutional Res. (2012)
5. Dietze, S., Sanchez-Alonso, S., Ebner, H., Qing Yu, H., Giordano, D., Marenzi, I., Pereira Nunes, B.: Interlinking educational resources and the web of data: a survey of challenges and approaches. Program 47, 60–91 (2013)
6. Daquin, M.: Linked data for open and distance learning. Common wealth of learning report (2014)
7. Zablith, F., d'Aquin, M., Brown, S., Green-Hughes, L.: Consuming linked data within a large educational organization. In: Presented at the Second International Workshop on Consuming Linked Data (COLD) at 10th International Semantic Web Conference (ISWC 2011), Bonn, Germany (2011)
8. Zablith, F., Fernandez, M., Rowe, M.: Production and consumption of university Linked Data. Interact. Learn. Environ. 23, 55–78 (2015)
9. Heath, T., Clarke, C., Singer, R., Leavesley, J., Shabir, N.: Assembling and applying an education graph based on learning resources in universities. In: Linked Learning (LILE) Workshop (2012)

10. Zablith, F.: Interconnecting and enriching higher education programs using linked data. In: Proceedings of the 24th International Conference on World Wide Web, pp. 711–716. International World Wide Web Conferences Steering Committee, Republic and Canton of Geneva, Switzerland (2015)

11. Rad, A., Naderi, B., Soltani, M.: Clustering and ranking university majors using data mining and AHP algorithms: a case study in Iran. Expert Syst. Appl. **38**, 755–763 (2011)

12. Cerbah, F.: RDBToOnto: un logiciel dédié à l'apprentissage d'ontologies à partir de bases de données relationnelles. In: proceeding of: Extraction etgestion des connaissances (EGC 2009), p. 495. Cépaduès, Strasbourg (2009)

13. Lahoud, I., Chamekh, F.: Towards an architecture for universities management. In: Presented at the Second International Workshop on Educational Knowledge Management co-located with 20th International Conference on Knowledge Engineering and Knowledge Management (EKAW 2016), Bologna, Italy (2016)

A Formalization of the French Elementary School Curricula

Oscar Rodríguez Rocha[1(✉)], Catherine Faron Zucker[2(✉)],
and Geraud Fokou Pelap[1(✉)]

[1] Inria Sophia Antipolis Méditerranée, Sophia Antipolis, France
{oscar.rodriguez-rocha,geraud.fokou-pelap}@inria.fr
[2] University Nice Sophia Antipolis, CNRS, I3S, UMR 7271,
Sophia Antipolis, France
faron@unice.fr

Abstract. In the education field, in order to achieve learning goals, it is necessary to define learning paths that foresee a gradual and incremental acquisition of certain knowledge and skills that students should acquire. In this paper we analyze the educational progressions of the French educational system, we show how to formalize them through a web ontology and how to perform knowledge extraction from the official texts describing them to automate the population of such an ontology.

Keywords: e-Education · Ontologies · Semantic web · Knowledge extraction · Knowledge representation

1 Introduction

In the education field, in order to achieve learning goals, it is necessary to define learning paths that foresee a gradual and incremental acquisition of certain knowledge and skills that students should acquire. Such educational programs are usually defined by a body (a ministry or department of education) and then implemented by the schools or institutes that are part of the educational system. Nowadays it is not possible to find them in a standardized format that can be accessible and queried automatically.

In this context, we propose an answer to the following research questions:

1. How to model educational programs into a formal machine understandable ontology that can be processed and queried?
2. How to extract the knowledge and skills that an educational program requires from the official texts describing them?
3. How to automate the population of the ontology based on the extracted knowledge?
4. How can we exploit these formal representations?

© Springer International Publishing AG 2017
P. Ciancarini et al. (Eds.): EKAW 2016 Satellite Events, LNAI 10180, pp. 82–94, 2017.
DOI: 10.1007/978-3-319-58694-6_7

Our scientific contribution is first, a web ontology that describes and represents an official standard of knowledge and skills, in this case the French educational system. It is formalized in the OWL Web Ontology Language, which enables to process and query it with SPARQL, the standard query language for the Web of data. Second, we propose a process to extract knowledge and skills from the official texts describing the French educational program. Finally, we propose a process to automatically populate our ontology with the knowledge extracted from the official texts which we further enrich with the web of data.

This work is a revised version of [10], published in the Proceedings of the Second International Workshop on Educational Knowledge Management co-located with the 20th International Conference on Knowledge Engineering and Knowledge Management (EKAW 2016)[8].

The remainder of this paper is structured as follows: Sect. 2 presents state of the art approaches of ontology-based modeling in e-Education. Section 3 introduces to the French common base of knowledge and skills for students as well as the official educational progressions. In Sect. 4 we propose EduProgression: an ontology to model the educational progressions of the French educational system. We show, in Sect. 5, how to exploit the data of the progressions modeled through EduProgression and its thesaurus. Section 6 presents our method to automate the population of this ontology by extracting elements of knowledge and skills from the official texts and Sect. 7 describes how to link them to relevant resources available on the Web of Data. Finally, the conclusions and future work are presented in Sect. 8.

2 Related Work

The term of "educational ontology" covers very different types of ontologies, modeling different kinds of knowledge related to e-Education. In [11], the authors provide a literature survey of the development and use of ontologies in the domain of e-learning systems. According to the classification proposed in this survey, our work falls within the "curriculum modeling and management" category. Among such works we have CURONTO [3,4], an ontological model designed for the management of a curriculum and to facilite program review and assessment". Compared to it, our ontology is not focused on giving the means to conduct a review of the program nor an assessment; it focuses on the description and representation of the knowledge and skills required by an educational system and it provides the means to link knowledge and skills with resources from the web of data.

In the same category, Gescur [6] is a tool dedicated to the management and evaluation of the implementation of the curriculum, which facilitates the curriculum management process". It relies on an ontology of concepts relevant to curriculum management in Secondary Schools, such as teachers, departments, objectives, subjects, modules, tasks, documents, policies, activities, learning objects, quality criteria, etc. While this ontology focuses on the administration of the curriculum ours focuses on the management of the knowledge and skills required in each school year in the French educational system. In this sense, both ontologies could be complementary.

Besides the above cited literature review, the educational semantic web ontology proposed in [5] focuses on representing higher education concepts and assisting specialized e-learning systems. In contrast to this work, we propose an ontology that focuses on representing the elements of knowledge and skills of an educational system.

Finally, the OntoEdu educational platform architecture for e-Learning proposed in [7], relies on an activity ontology that describes all the education activities and the relations among them, and a material ontology which describes the educational content organization. The ontology is not available online, therefore it is not possible neither to analyze its classes and properties nor to determine the feasibility of use it in other scenarios outside the OntoEdu platform. Additionally their ontology does not contemplate the use of resources of the web of data to enrich the educational material.

3   The French Educational Model

In France, there is an official common base of knowledge and skills created by the Ministry of Education and published in the Official Journal; it is called the *"Socle commun de connaissances et de compétences"*[1]. It states that, compulsory education should at least guarantee to each student the necessary means to acquire a common base consisting of a set of knowledge and skills that are essential to master in order to successfully complete her education, build her personal and professional future and succeed in her life in society. The current common base of knowledge and skills entered into effect in 2016 while its predecessor leasted from 2006 to 2016. Our work was originally based on the first version of the French common base, but as we will show in this paper, it has been adapted to be compatible with the new version.

The French common base of 2006 was organized according to seven skills:

- Mastery of the French language
- Practice of a foreign language
- Basic math skills and scientific and technological culture
- Mastery of the common information and communication techniques
- Humanistic culture
- Social and civic competences
- Autonomy and initiative of the students

On the other hand, the current French common base is made up of five different skill domains:

- Language for communicate and thinking
- Methods and tools for studying
- Formation to the person and citizen

---

[1] http://www.legifrance.gouv.fr/WAspad/UnTexteDeJorf?numjo=MENE0601554D.

- Naturals systems and technical systems
- Representation of the world and human activities

These skill domains represent the knowledge and skills that students must acquire for a particular and essential requirement of the life in society.

In order to organize a gradual acquisition of the knowledge and skills of the French common base, some official learning progressions *"Progressions pour l'ecole elmentaire"* for the educational areas of the French elementary school have been defined and published (official bulletin of January 5, 2012). In the latest French common base, such progressions have been integrated implicitly into the division of the learning of each domain as three different cycles [2] with one precise goal for each of them. Thus, we have:

- Cycle 2 (CP-CE1-CE2): learning fundamentals
- Cycle 3 (CM1-CM2-6^{ieme}): consolidation
- Cycle 4 (5^{ieme}-4^{ieme}-3^{ieme}): deepening

These cycles were already defined in the common base of 2006, but used for a different goal. For each elementary school cycle, the progressions are organized by area of education and they strictly follow the program labels to offer a comprehensive approach to knowledge, skills and attitudes to be mastered by the students. In this common base, the progressions are focused on a skill while in the common base of 2016 the progression are focused on a learning domain, that means on all the skills needed for a particular learning domain.

Our study is based on the analysis of these progressions (which are based on the French common base of knowledge and skills). By generalizing them, we built a standard model of the knowledge and skills necessary for the students through their educational path and a method to automate the population of this model by extracting knowledge from the texts describing these progressions. Being a standard model, it is also possible to model progressions from other educational systems in the world, such as the U.S. academic standards[2] defined by the Common Core State Standards (CCSS)[3].

4   The EduProgression Ontology and Associated thesaurus

4.1   The EduProgression Ontology

We created EduProgression, an ontological model formalized in the standard Ontology Web Language (OWL) for the formalization of educational progressions or programs as defined in the French common bases of 2006 [1] and 2016. In this formalization, we identified the following main classes:

---

[2] http://www.corestandards.org/read-the-standards/.
[3] http://www.corestandards.org/.

- **Element of Knowledge and Skills (EKS).** As skills and knowledge are keystones of the common bases, this element is the key concept of our model. It is represented by the EKS class, which is the central class of the EduProgression ontology. This class is common for the two releases of the common base.

  An element of knowledge and skills is associated to a set of knowledge pieces (class `Knowledge`) and/or skills (class `Skill`) for a specific French school cycle (class `Cycle`) or course (class `Course`) that may contain reference points (class `PointOfReference`) and also a vocabulary of associated terms (class `Vocabulary`). More precisely:

  - `Knowledge`. Instances of class `Knowledge` are also instances of `skos:Concept` and each one belongs to a `skos:ConceptScheme` that contains all the knowledge pieces of a given progression. An instance of `EKS` is related to an instance of `Knowledge` through the property `hasKnowledge`.
  - `Skill`. An instance of `EKS` is related to an instance of `Skill` through property `hasSkill`.
  - `Course`. In the French common base of 2006, skills that students are expected to develop, are defined by cycle and each cycle is organized into courses. For example, "the consolidation cycle" includes "the second year elementary course" (CE2), "the first-year intermediate course" (CM1) and "the second year intermediate course" (CM2). In this context, an instance of class `Course` represents a course in a cycle. An instance of `EKS` is related to an instance of `Course` through property `hasCourse`.
  - `Cycle`. In the French common base of 2016, skills that students are expected to develop, are defined only by cycle. An instance of `EKS` is related to an instance of `Cycle` through property `hasCycle`.
  - `PointOfReference`. An instance of class `PointOfReference` represents an educational reference element on a specific element of knowledge and skills (an instance of `EKS`). An instance of `EKS` is related to an instance of class `PointOfReference` through property `hasPointOfReference`.
  - `VocabularyItem`. Each element of knowledge and skills has vocabulary items. This vocabulary is compatible for the two common bases. An instance of `EKS` is related to an instance of class `VocabularyItem` through property `hasVocabularyItem`. A vocabulary item is also an instance of `skos:Concept` and it is related to an instance of `skos:ConceptScheme` which gathers the concepts of the thesaurus of the progression.

- **Progression.** In the common base released in 2006, the progressive acquisition of knowledge and skills is defined as a "progression", while for the current common base, the progressive acquisition of skills is defined as a "program". Therefore, a progression or a program in our model, is represented as an instance of the class `Progression`. It can be associated to an existing learning domain (through property `hasLearningDomain`) and to one or several EKSs (through property `hasEKS`).

- **Learning domain.** A learning domain represents a school subject like *History* or *Mathematics*. The learning domain is represented, in the ontology EduProgression, by an instance of the `LearningDomain` class, and it is also an instance of `skos:Concept` that is part of (only) one `skos:ConceptScheme` containing the only learning domains of a progression. Also, as they are SKOS concepts, learning domains are hierarchically organized by using the `skos:broader` and/or `skos:narrower` properties. A learning domain can be associated to a `Progression` or an EKS.

- **Skills Domain.** For the common base of 2016, each domain of skills can be represented in our model by a class named `SkillsDomain`. An EKS can be associated to one or many skills domains through the property `hasSkillsDomain` to represent the skills domain(s) that it targets.

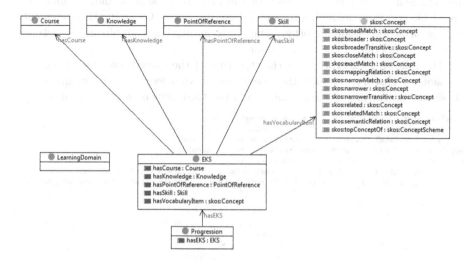

Fig. 1. Classes and properties of the EduProgression ontology

The EduProgression ontology is freely accessible online.[4] It is composed of 7 OWL classes and 8 OWL object properties. The relationships between them can be seen in Fig. 1. This figure presents the ontology for the education system released in 2006.

The two following excerpt is the description of the History progression and a related *EKS* representing the "european construction" based on EduProgression. We highlight with three different colors the three different types of resources related to specific thesaurus described in the following (see Sect. 4.2):

```
@prefix edp: <http://ns.inria.fr/semed/eduprogression> .
edp:Histoire a edp:Progression ; rdfs:label "History"@en, "Histoire"@fr ;
    edp:hasEKS edp:La_construction_europeenne .
edp:La_construction_europeenne a edp:EKS ;
    rdfs:label "The european construction"@en ;
    rdfs:label "La construction européenne"@fr ;
    edp:hasKnowledge edp:Le_role_de_Robert_Schuman ;
    edp:hasLearningDomain edp:Le_XXe_siecle_et_notre_epoque ;
```

---

[4] http://ns.inria.fr/semed/eduprogression.

```
edp:hasVocabularyItem edp:Euro .
edp:hasCourse edp:CM2 ;
edp:hasPointOfReference edp:Euro_monnaie_europeenne ;
edp:hasRelatedResource dbpedia-fr:Union_européenne ;
edp:hasSkill edp:Principales_etapes_de_la_construction_europeenne ;
```

4.2   Three Thesaurus Associated to Each Educational progression

Each formalized progression is associated to three thesaurus: one for its knowledge pieces, one for its vocabulary items and another one containing its learning domains. As a result, *knowledge pieces, vocabulary items* and *learning domains* are represented by instance classes of the EduProgression ontology which also are SKOS concepts belonging to one of these thesaurus. Grouping these elements in thesaurus allows, among other benefits, to manage in a simple and structured way the knowledge of each progression [9].

The following excerpt shows the structure of the three thesaurus of the progression of History (in blue the thesaurus of knowledge pieces, in green the thesaurus of learning domains and in red the thesaurus of vocabulary items).

```
edp:ElementsConnaissanceHistoire a skos:ConceptScheme ;
    skos:prefLabel "Knowledge pieces of History"@en .
    skos:prefLabel "Elements de connaissance de l'Histoire"@fr .
edp:Le_role_de_Robert_Schuman a edp:Knowledge, skos:Concept ;
    skos:inScheme edp:ElementsConnaissanceHistoire ;
    skos:prefLabel "Know the role of Robert Schuman in European construction"@en .
    skos:prefLabel "Connaître le rôle de Robert Schuman dans la construction européenne"@fr
                 .
edp:DomainesApprentissageHistoire a skos:ConceptScheme ;
    skos:prefLabel "Learning domains of History"@en .
    skos:prefLabel "Domaines d'apprentissage de l'Histoire"@fr .
edp:Le_XXe_siecle_et_notre_epoque a edp:LearningDomain, skos:Concept ;
    skos:inScheme edp:DomainesApprentissageHistoire ;
    skos:prefLabel "Elements of knowledge and skills on the twentieth century and our
                 time"@en ;
    skos:prefLabel "Éléments de connaissances et de compétences sur le XXe siécle et notre
                 époque"@fr .
edp:VocabulaireHistoire a skos:ConceptScheme ;
    skos:prefLabel "Vocabulary of History"@en ;
    skos:prefLabel "Vocabulaire de l'Histoire"@fr .
edp:Euro a skos:Concept ;
    skos:inScheme edp:VocabulaireHistoire ;
    skos:prefLabel "Euro"@en, "Euro"@fr .
```

To give a clearer view of the elements of the History progression described in the previous excerpts, Fig. 2 presents it as a graph.

At the time of writing, we have created the thesaurus of the vocabulary, knowledge pieces and learning domains relative to three progressions from the French curricula: History, Geography and Experimental sciences and technology, and we have populated the EduProgression ontology with the concepts in these thesaurus. For the progression on History, we identified 29 elements of knowledge and skills (instances of edp:EKS), the thesaurus of vocabulary items comprises 131 concepts, the thesaurus of knowledge pieces 45 concepts and the thesaurus of learning domains 7 concepts. For the progression of Geography, 32 elements of knowledge and skills were identified; its thesaurus of vocabulary items is composed of 131 concepts; its thesaurus of knowledge pieces and learning domains

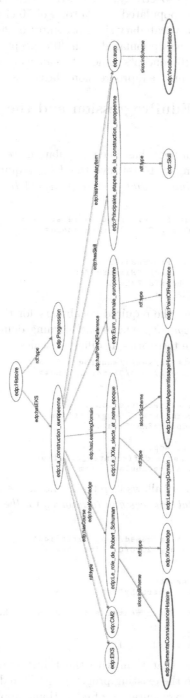

Fig. 2. Graph of the History progression excerpt

contains 34 and 6 concepts respectively. Finally, the pogression of Experimental sciences and technology was populated with a total of 50 elements of knowledge and skills; It has the largest vocabulary since it comprises 265 elements. Its thesaurus of knowledge pieces was populated with 40 concepts while its thesaurus of learning domains with 10 concepts. These thesaurus can be freely accessed online: http://ns.inria.fr/semed/eduprogression/thesaurus.

5   Exploitation of EduProgression and the Associated thesaurus

As the EduProgression ontology has been populated, we can exploit it with SPARQL queries. For example, to retrieve the *EKS* required for the students attending the *second year intermediate course* (CM2) of *History*, we can use the following query:

```
PREFIX edp: <http://ns.inria.fr/semed/eduprogression#>
PREFIX rdfs: <http://www.w3.org/2000/01/rdf-schema#>
SELECT ?eks
WHERE {
  ?history_progression a edp:Progression .
  ?history_progression rdfs:label "History"@en .
  ?history_progression edp:hasEKS ?eks .
  ?eks edp:hasCourse edp:CM2 . }
```

Similarly, we can retrieve the required vocabulary for *the first-year intermediate course* (CM1) of *Geography*, within the learning domain of "Elements of knowledge and skills on the territories at different scales".

```
PREFIX edp: <http://ns.inria.fr/semed/eduprogression#>
PREFIX rdfs: <http://www.w3.org/2000/01/rdf-schema#>
SELECT ?vocabulary_item
WHERE {
  ?geography_progression a edp:Progression .
  ?geography_progression rdfs:label "Geography"@en .
  ?geography_progression edp:hasEKS ?eks .
  ?eks edp:hasCourse edp:CM1 .
  ?eks edp:hasLearningDomain edp:
      ECC_sur_les_territoires_a_differentes_echelles .
  ?eks edp:hasVocabularyItem ?vocabulary_item . }
```

Finally, the following query allows to retrieve the necessary knowledge pieces of *History* and *Experimental sciences and technology* for *the second year elementary course* (CE2).

```
PREFIX edp: <http://ns.inria.fr/semed/eduprogression#>
PREFIX rdfs: <http://www.w3.org/2000/01/rdf-schema#>
SELECT ?knowledge
WHERE {
  ?progression a edp:Progression .
  {?progression rdfs:label "Geography"@en }
  UNION
  {?progression rdfs:label "Experimental sciences and technology"@en }
  ?progression edp:hasEKS ?eks .
  ?eks edp:hasCourse edp:CE2 .
  ?eks edp:hasKnowledge ?knowledge . }
```

It is important to note that thanks to the EduProgression ontology it is possible to make queries to the existing progressions filtered by specific *courses*. This is the basis for the implementation of educational applications adapted to each student profile.

6 Knowledge Extraction from Texts and Formalization

For the extraction of the information contained in the progressions for elementary school we had to consider first, that the progressions are only available in PDF format, that is, a PDF document for each learning domain. The data contained in each PDF, was inside tables without a standard format, therefore, when we tried to use common techniques for PDF data extraction[5] such as OCR or data extraction from tables[6], the resulting extracted text was mixed among various tables and/or incomplete.

This first led us to perform a preliminary manual extraction of knowledge from text. The first process that we experimented aimed at directly extracting RDF data from PDF texts. This involved to copy textual data directly from each PDF file, and then to paste it into an RDF description as literal values of some RDF triples (also manually written). Not only this process was time-consuming but also it was prone to human errors in the format of the data and the resulting RDF triples.

To solve these problems, we split the data extraction process in two steps, the first one consisting in manually filling a JSON intermediate data structure, the second one consisting in automatically transforming JSON data into RDF data. We designed an intermediate data structure in JSON format in which we represented each progression as a JSON object containing the following information:

- The name of the progression
- An array of the different learning sub domains involved in the progression
- A dictionary with the French scholar courses covered
- An array of the elements of knowledge and skills (EKS) involved in the progression. Each EKS itself contains:
  - a title and a description in French
  - the scholar cycle covered by this EKS
  - the learning domain covered by this EKS
  - a dictionary of the knowledge pieces (titles and descriptions in French) required by this EKS
  - a dictionary containing the required skills (titles and descriptions in French) required by this EKS
  - an array of the vocabulary items required by this EKS
  - an array of points of reference required by this EKS

This allowed us to manually fill the data structure in a much faster and much more reliable way. In addition, in order to validate the format of the data, we defined a JSON schema[7] representing these aforementioned structural constraints. This enabled us to automatically validated the JSON objects representing a progression manually entered.

---

[5] https://en.wikipedia.org/wiki/Optical_character_recognition.
[6] http://tabula.technology/.
[7] http://json-schema.org/.

As a second step, we defined and performed an automatic conversion process of JSON objects into RDF graphs. It uses the EduProgression ontology to automatically generate an RDF graph.

The following excerpt is the intermediate JSON representation of the EKS on the Age of Discovery, which conversion in RDF is presented in the excerpt of Sect. 4.2, which illustrates the use of the EduProgression ontology.

7   Entity Linking

We enriched the EduProgression ontology and its related thesaurus by conducting an entity linking process. It allows us to link the EKS instances of our already populated progressions, with DBpedia resources related to the textual values of the properties describing EKS instances. The information provided by these linked resources may be further exploited directly as learning material or to discover related learning resources.

To make this possible, we have created an automated process that first extracts literal values of the properties of each EKS and builds a textual representation of the latter by concatenating these literal values, and then retrieves the related DBpedia resources by using the DBpedia spotlight web service. The process is as follows:

1. All the instances of *EKS* are retrieved from the populated progressions. This is the result of a SELECT *SPARQL* query processed by a **SPARQL engine** on our local triple store.
2. For each retrieved *EKS* instance, the textual values of its properties are extracted and concatenated into a string which will be considered as a whole as the **EKS text representation**.
3. The process takes each **EKS text representation** and analyzes it by using the DBPedia Spotlight web service.
4. Each related resource obtained from the previous step is linked to the *EKS* by using the `edp:hasRelatedResource` property of EduProgression. This is done with an INSERT *SPARQL* query applied to our triple store.

The whole process is shown in Fig. 3.

Currently, our process creates links between EKS instances and related resources in DBpedia. We chose to link EKS since they are conceived in our model as the main concept that represents skills and knowledge. Eventually, for other case studies, the EduProgression ontology could be adapted in such a way that the property `hasRelatedResource` can be applied to other types of instances, among which skills, knowledge pieces and vocabulary items, making it possible to link their instances with related resources from the web of data.

As a result of our process, we have obtained 187 resources related to the EKSs of the Geography progression with a precision of 94%, 205 for the History progression with a precision of 95% and 220 for the progression of Experimental Sciences and technology with 94% of precision. For these experiments DBpedia was configured with a confidence of 0.15 and a support of 20.

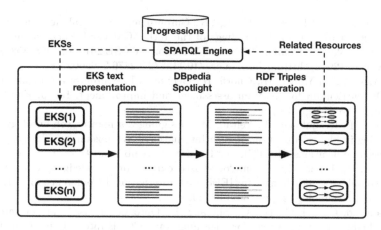

Fig. 3. Process to link EKS instances of a given progression to DBpedia resources

8   Conclusions and Future Work

In this paper, we have shown how to model an educational program by formalizing it as a Web ontology in OWL that is understandable by machines and that can be queried and processed. In particular, we have taken as a case study the French education system, from which we have extracted and formalized educational progressions in an ontology called EduProgression. We sketched how this ontology can be used through some example SPARQL queries. We have described a process for populating the EduProgression ontology, by extracting the knowledge elements and skills involved in the educational progressions from the official texts describing these progressions. Additionally, we showed how we managed to automatically link the ontology instances to Web resources.

As stated before, our work was originally based on the French "Socle commun de connaissances et de compétences" of 2006, however, as an update to this common base has been made in September of 2016, we have adapted our work accordingly, maintaining also the compatibility with the previous common base.

Future work includes the use of the EduProgression ontology and its related thesaurus for the indexation of educational resources to the French curricula through our ontology, with the aim of recommending them to students in their learning path. This activity will be performed in collaboration with our two industrial partners. In addition, we look forward to apply this work to other educational systems in the world as suggested in Sect. 3.

References

1. Le socle commun de connaissances et de compétences. http://media.education. gouv.fr/file/46/7/5467.pdf
2. Programmes d'enseignements. http://cache.media.education.gouv.fr/file/MEN_ SPE_11/67/3/2015_programmes_cycles234_4_12_ok_508673.pdf

3. Al-Yahya, M., Al-Faries, A., George, R.: Curonto: an ontological model for curriculum representation. In: Proceedings of the 18th ACM Conference on Innovation and Technology in Computer Science Education, ITiCSE 2013, pp. 358–358. ACM, New York (2013). http://doi.acm.org/10.1145/2462476.2465602

4. Alfaries, A., Al-Yahya, M., Chorfi, H., George, R.P.: Curonto: a semantic model of the curriculum for program assessment and improvement. Int. J. Eng. Educ. **30**(5), 1083–1094 (2014)

5. Bucos, M., Dragulescu, B., Veltan, M.: Designing a semantic web ontology for e-learning in higher education. In: 9th International Symposium on Electronics and Telecommunications (ISETC), pp. 415–418, November 2010

6. Dexter, H., Davies, I.: An ontology-based curriculum knowledgebase for managing complexity and change. In: IEEE 14th International Conference on Advanced Learning Technologies, pp. 136–140 (2009)

7. Guangzuo, C., Fei, C., Hu, C., Shufang, L.: Ontoedu: a case study of ontology-based education grid system for e-learning. In: GCCCE International Conference, Hong Kong (2004)

8. Lahoud, I., Matta, N., Zablith, F., Ventura, S. (eds.) Proceedings of the Second International Workshop on Educational Knowledge Management co-located with 20th International Conference on Knowledge Engineering and Knowledge Management (EKAW 2016), Bologna, Italy, 19–23 November 2016, CEUR Workshop Proceedings, vol. 1780 (2017). CEUR-WS.org, http://ceur-ws.org/Vol-1780

9. Pastor-Sánchez, J.A., Martínez Méndez, F.J., Rodríguez-Muñoz, J.V.: Advantages of thesaurus representation using the simple knowledge organization system (SKOS) compared with proposed alternatives. Inf. Res. Int. Electron. J. **14**(4), n4 (2009)

10. Rocha, O.R., Faron-Zucker, C.: A formalization of standard knowledge and skills for the french elementary school curricula. In: Lahoud et al.: [8], pp. 60–71. http://ceur-ws.org/Vol-1780/paper6.pdf

11. Yahya, M.A., George, R., Alfaries, A.: Ontologies in e-learning: review of the literature. Int. J. Softw. Eng. Appl. **9**(2), 67–84 (2015)

Posters and Demos

Semantic Integration of Geospatial Data from Earth Observations

Helbert Arenas[(✉)], Nathalie Aussenac-Gilles[iD], Catherine Comparot,
and Cassia Trojahn

Institut de Recherche en Informatique de Toulouse, Toulouse, France
{helbert.arenas,nathalie.aussenac-gilles,catherine.comparot,
cassia.trojahn}@irit.fr

Abstract. We propose an approach to semantically enrich metadata records of satellite imagery with external data. As a result we are able the identify relevant images using a larger set of matching criteria. Conventional methods for annotating data sets are usually based on metadata records (with attributes such as title, provider, access mode, and spatio-temporal characteristics), which offer a narrow view of the world. Enriching metadata with contextual information (e.g., the region depicted in the image has been recently affected by extreme weather) requires formalizing spatio-temporal relationships between metadata records and external data sources. Semantic technologies play a key role in such scenarios by providing an infrastructure based on RDF and ontologies.

Keywords: Earth observation · Geospatial semantics

1 Introduction

Traditional approaches for the description and identification of satellite imagery make use of metadata records. However they are limited in the sense that they consider metadata records as isolated entities. Our work contributes to Geospatial semantics [3] enabling interoperability between data sources. We propose to use open data sources and ontologies to define formal relationships between metadata records and external data sources. Thus we are able to create links to useful pieces of information in the image spatio-temporal proximity. This new information provides the image with context, enhancing its description.

In order to cover our annotation needs (image metadata, earth observation, sensors, and specific contextual data, with special attention to spatio-temporal data), we use three standard vocabularies: (1) Data Catalog Vocabulary (DCAT), a W3C standard designed to handle catalog records on the web [5]; (2) Semantic Sensor Network (SSN) another W3C standard designed to describe sensors and observations [1]; and (3) GeoSPARQL [4], an OGC standard designed as a minimum ontology to query geospatial data on the Semantic Web.

This paper details the proposed model for spatial and temporal data integration. We illustrate our approach using Sentinel images and weather information as an external data source. We use GeoSPARQL to dynamically create the links between the metadata records and their contextual information.

P. Ciancarini et al. (Eds.): EKAW 2016 Satellite Events, LNAI 10180, pp. 97–100, 2017.
DOI: 10.1007/978-3-319-58694-6_8

2   Data Sets

This work is carried out within the SparkInData[1] project, thus we use metadata records of Sentinel images[2]. The revisit time for Sentinel 1 is twelve days, while for Sentinel 2 is five days. The metadata records are obtained from RESTO, a data service managed by CNES (Centre National d'Etudes Spatiales) [2] in JSON format. We have developed a software tool to translate the records to RDF, it uses the Jena and GeoTools libraries.

As a contextual data we use weather information provided by *Meteo France*[3]. This organization offers data as monthly compiled CSV zipped files. The observations are taken every three hours for each one of the 62 weather stations in France. A separate file contains a list of the weather stations with their position as points encoded as geographic coordinates. We have developed an application in Java to map the information contained in the CSV files into our Data Model as RDF.

3   A Data Model for Enriched Geospatial Metadata Records

Figure 1 depicts the proposed data model. The class *SatelliteImgMetadataRecord* is a subclass of *dcat:Dataset*. This specialization is required in order to represent properties specific of satellite imagery, such as the platform (what Satellite is taking the image), the processing level, the sensor mode (which depends on the satellite and the observed surface). A satellite image metadata record has a spatial dimension and a temporal dimension. Both of these dimensions are used to add external data to the image.

The model handles the **spatial dimension** of a metadata record thanks to the property *hasSpatialRepresentation* and the GeoSPARQL ontology. This property has as range the class *geo:Feature* and as domain the class *SatelliteImgMetadataRecord*. Thanks to this property we are able to use GeoSPARQL functions to relate the metadata records with any other information with a spatial component and also defined as a *geo:Feature*. GeoSPARQL provides formal definitions of topological relationships (RCC8). Triplestores that support GeoSPARQL, also offer a set of spatial functions that enable geospatial operations, such as *geof:distance, geof:intersection, geof:difference*, etc. By using GeoSPARQL, we can create statements that describe the topological relationships between a satellite image footprint and other entities with a spatial nature.

In order to provide contextual information to the metadata records we link them to weather information that we model using both the Semantic Sensor Network (SSN) and the GeoSPARQL vocabularies. We represent a weather station as an instance of the class *MeteoFranceWeatherStation*, which is at the same time a subclass of *ssn:SensingDevice* and as a subclass of *geo:Feature*. Then it is

---

[1] SparkInData project is funded by "Investing for the Future" French program.

[2] https://sentinel.esa.int/web/sentinel/missions/ (07/2016).

[3] https://donneespubliques.meteofrance.fr/ (07/2016).

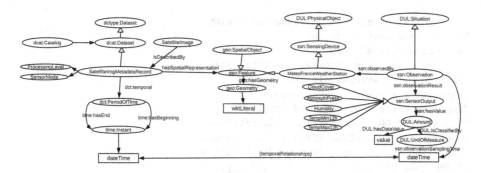

Fig. 1. Main classes used in our model.

possible to compare the two instances of the class *geo:Feature* and to determine their spatial relationships.

The *temporal dimension* of an image metadata record identifies the moment when the image has been captured. The external data source, the weather information, also has a temporal dimension. Weather stations record measurements periodically. So we use the class *ssn:Observation* as a way to relate the measured variables to the weather station while at the same time, providing a temporal dimension for the observations (Fig. 1). Then, it is possible to establish temporal relationships (before, after) between an image metadata record and weather measurements or to establish periods of interest (e.g., one week after the image was created). Using the model, the user can define a relevant period of time and link an image metadata record to the available weather information. The user defined period works as a temporal buffer that provides context to the image metadata record.

The following code depicts a SPARQL query that links image metadata records with weather measurements corresponding to stations located inside the image footprint (cf. *FILTER(geof:sfContains(?img_md_geo, ?ws_geo)))* and the data collected during a given period. The user defined a period of interest of one week after the image was taken ($24 \times 7 = 168$ h, 'PT168H') that is used to filter the relevant data (cf. the last three FILTER lines)[4]. We have used Stardog[5] (version 4.0.5) as triplestore supporting GeoSPARQL.

```
select ?img_md ?img_identifier ?obs_time ?humidityDataVal ?humidityUnits
where{
?img_md a sparkindata:SatelliteImgMetadataRecord.
?img_md dct:identifier ?img_identifier.
?img_md sparkindata:hasSpatialRepresentation ?img_md_feat.
?img_md_feat geo:hasGeometry ?img_md_geo.
?img_md dct:temporal ?img_md_temp.
?img_md_temp time:hasEnd ?img_md_end.
```

---

[4] A video showing the implementation of our approach is available at: http:// geo-space.info/demos/metadataBrowserDataMeteoFrance.mp4.

[5] http://stardog.com/.

```
?ws a sparkindata:MeteoFranceWeatherStation.
?ws geo:hasGeometry ?ws_geo.
?obs ssn:observedBy ?ws.
?obs ssn:observationSamplingTime ?obs_time.
?obs ssn:observationResult ?humidity.
?humidity a sparkindata:Humidity.
?humidity ssh:hasValue ?humidityVal.
?humidityVal DUL:hasDataValue ?humidityDataVal.
?humidityVal DUL:isClassifiedBy ?humidityUnits.
BIND (?img_md_end-?obs_time as ?diffDateTime)
FILTER(geof:sfContains(?img_md_geo,?ws_geo))
FILTER (?diffDateTime<'PT168H'^^xsd:dayTimeDuration)
FILTER (?diffDateTime>'PT0H'^^xsd:dayTimeDuration)}
```

4   Concluding Remarks and Future Work

We have presented an approach for semantic annotation of satellite images that relies on a model providing context to image metadata records by using external data sources. The proposed approach is rather generic and is based on well-known existing vocabularies. We show the feasibility of this approach by using weather information as the contextual data. Currently, we are working on extending this approach to use as contextual information phytosanitary reports and seismic information. As future work, we plan to integrate automatic matching approaches for matching raw image metadata and vocabularies as well as to align semantic annotated data with other sources of data in the Linked Open Data. We plan as well to consider domain-oriented sources of data for a particular use case (agriculture and data sources as agricultural reports) and provide rules and reasoning capabilities that help the specific domain analysis.

References

1. Compton, M., Barnaghi, P., Bermudez, L., García-Castro, R., Corcho, O., Cox, S., Graybeal, J., Hauswirth, M., Henson, C., Herzog, A., Huang, V., Janowicz, K., Kelsey, W.D., Le Phuoc, D., Lefort, L., Leggieri, M., Neuhaus, H., Nikolov, A., Page, K., Passant, A., Sheth, A., Taylor, K.: The SSN ontology of the W3C semantic sensor network incubator group. J. Web Seman. **17**, 25–32 (2012)
2. Gasperi, J.: Semantic search within earth observation products database based on automatic tagging of image content. In: Proceedings of the Conference on Big Data from Space, pp. 4–6 (2014)
3. Janowicz, K., Scheider, S., Pehle, T., Hart, G.: Geospatial semantics and linked spatiotemporal data -past, present, and future. Seman. Web **3**(4), 321–332 (2012)
4. Kolas, D., Perry, M., Herring, J.: Getting started with GeoSPARQL. Technical report, OGC (2013)
5. Maali, F., Erickson, J.: Data Catalog Vocabulary (DCAT) (2014)

Matching Ontologies
Using a Frame-Driven Approach

Luigi Asprino[1,2]([⊠]), Valentina Presutti[2], and Aldo Gangemi[2,3]

[1] DISI, University of Bologna, Bologna, Italy
luigi.asprino@unibo.it
[2] Institute of Cognitive Sciences and Technologies,
National Research Council of Italy, Rome, Italy
[3] LIPN, Université Paris 13 Sorbonne Cité, CNRS, Paris, France
{valentina.presutti,aldo.gangemi}@cnr.it

Abstract. The need of handling semantic heterogeneity of resources is a key problem of the Semantic Web. State of the art techniques for ontology matching are the key technology for addressing this issue. However, they only partially exploit the natural language descriptions of ontology entities and they are mostly unable to find correspondences between entities having different logical types (e.g. mapping properties to classes). We introduce a novel approach aimed at finding correspondences between ontology entities according to the intensional meaning of their models, hence abstracting from their logical types. Lexical linked open data and frame semantics play a crucial role in this proposal. We argue that this approach may lead to a step ahead in the state of the art of ontology matching, and positively affect related applications such as question answering and knowledge reconciliation.

1 Introduction

Ontologies are artifacts encoding a description of a domain of interest for some purpose. Due to the Web's open nature, ontologies can be defined by different people and can vary in quality, expressiveness, richness, and coverage, hence increasing semantic heterogeneity of the resources made available through the Web of Data. Among the various semantic technology proposed to handle heterogeneity Ontology Matching [9] has proved to be an effective solution to automate integration of distributed information sources. Ontology Matching (OM) finds correspondences between semantically related entities of ontologies. However, most of the current ontology matching solutions present two main limits: (i) they only partially exploit the natural language descriptions of ontology entities and lexical resources as background knowledge; (ii) they are mostly unable to find correspondences between entities specified through different logical types (e.g. mapping properties to classes).

Frame Semantics [2] is a formal theory of meaning based on the idea that human can better understand the meaning of a single word by knowing the relational knowledge associated to that word. For example, the meaning of the verb *buy* can be clarified by knowing that it is used in a situation of a commercial

© Springer International Publishing AG 2017
P. Ciancarini et al. (Eds.): EKAW 2016 Satellite Events, LNAI 10180, pp. 101–104, 2017.
DOI: 10.1007/978-3-319-58694-6_9

transfer which involves individuals playing specific roles, e.g. a buyer, a seller, goods, money and so on. In other words, the verb buy *evokes* a scene where there are some individuals are playing specific roles. Our hypothesis is that the frames evoked by words associated with an ontological entity can be used to derive the intended meaning of that entity thus facilitating the ontology matching task.

In this paper we introduce a novel approach aimed at finding correspondences between ontology entities according to the intensional meaning of their models, hence abstracting from their logical types. This strategy allows us to match ontological entities with respect to their intensional meaning (that we suppose is evoked by the textual annotations associated with them) instead of their axiomatization, hence to abstract from their logical type. In fact, the axiomatization could have been forced by the choice of certain language for specifying the ontology, by the personal modeling style of the designer, or, other requirements (e.g. the compatibility with an existing ontology) unrelated to the modeled domain.

2   Proposed Approach

Following [4], we devised an approach for ontology matching that considers frames as *"unit of meaning"* for ontologies and exploits them as means for representing the intensional meaning of the entities. Our strategy consists of three steps, summarized as follows.

Selecting frames evoked by annotations. In order to associate ontological entities with frames we analyze the textual annotation associated with them. Annotations provide humans with insights of the intensional meaning the designer wants to represent with a certain entity. The main idea of this approach is that words used in annotations *evoke* frames that are representative of the intensional meaning of the entity. In associating entity with frames, the ambiguity of words has to be taken into account. For instance, the verb *bind* evokes either the FrameNet's frame *Imposing obligation* or *Becoming attached*. Therefore, to associate entities with the most appropriate frames, we have: (i) to associate words in the entities' annotation with the most appropriate sense (WSD by using UKB [1] and Babelfy [5]); (ii) and then, to select evoked frames by exploiting the Framester's mapping between WordNet's synsets and FrameNet's frames [3]. This approach is able to associate ontology entities to frames even if its annotations use specialized terminology. In this case it is exploited the Framester's mapping from Babelnet synsets and DBPedia resources[1] to frames. At the end of this step ontology entities are associated with a set of frames. For instance the object property `isParticipantIn` of the ODP Participation[2] is associated with the frames: *Participation, Collaboration, People* and *Evaluative comparison*.

Mapping frames and ontologies. This step creates an effective mapping between ontology entities and frames evoked by its textual annotations. An

---

[1] Both Babelfy and UKB are able to perform entity linking over text.
[2] http://ontologydesignpatterns.org/wiki/Submissions:Participation.

example of mapping is provided by FrameBase's integration rules [8]. In order to identify the effective mapping between ontologies and frames, for each entity we compute any possible mapping between the entity and the frames selected in the previous step (i.e. those evoked by its annotations). In frame semantics, a frame is characterized by its roles (also called frame elements) and each element possibly define the semantic type of the individual that can play that role in the frame. Frames, frame elements and semantic types have a name and a description. For each ontology entity we compute the semantic text similarity (by means of ADW [6]) between the textual annotations of the ontology entity and those associated with the evoked frames, its elements, and its semantic types. We map the ontology entity to the top-scoring frame entity in semantic text similarity. For instance, it easy to see that the top-scoring alignment for `isParticipantIn` is that mapping it on the frame *Participation*[3], its domain/range (i.e. `Object` and `Event`) on the frame elements *Participant* and *Event*, respectively.

Frame-based ontology matching. Once input ontologies and frames are aligned, each ontology entity is associated with a formal specification of its intensional meaning (that we call *frame-based specification*). As pointed out in [7] the properties *subclass of* and *sub-property of* are not enough to explicit complex relation between entities. In light of this consideration we express the relation between frames and ontology entities by interpreting both as *predicates*. A formalization of frames as *multigrade predicates* is provided by [3]. A straightforward interpretation of ontology entities as predicates represents classes as n-ary predicates (the arguments of the n-ary predicate are the entities in its neighborhood) and properties as binary predicates. For instance, the class `TimeIndexedPartipation`[4] can be represented as a ternary predicate with arguments provided by `Event`, `TemporalEntity` and `Object`. Interpreting frames and ontology entities in predicates allows us to express complex relationship which cannot be formalized by only using OWL/RDFs vocabularies. Framester ontology [3] defines a set relationship holding between predicates. Using the Framester vocabulary the class `TimeIndexedPartipation` can be specified as `projectionOf` the frame *Participation*, with members `involveEvent`, `atTime` and `includesObject` (which can be interpreted as subroles of *Event*, *Time* and *Participant*). Also the property `isParticipantIn` of the ODP *Participation* can be specified as `projectionOf` the frame *Participation*, with members `Object` and `Event`. Therefore, the class `TimeIndexedParticipation` and the object property `isParticipantIn` are "aligned" to the same frame and a complex correspondence between `TimeIndexedParticipation` and `isParticipantIn` can be derived. In this case `isParticipantIn` is a `subframeOf` `TimeIndexedParticipation`. The subframe relation might be used for creating a CONSTRUCT SPARQL query or an inference rule[5] transforming instances of the class in instances of the property.

---

[3] FrameNet Frame Participation https://goo.gl/IMdAwA.
[4] Time Indexed Participation ODP https://goo.gl/qX3DDr.
[5] Refer to [8] for examples of these kinds of rules.

3   Conclusion and Future Work

In this paper we introduced a novel approach for ontology matching. This method exploits the frame semantics as cognitive model for representing the intensional meaning of ontology entities. The frame-based representation enabled finding correspondences between ontology entities abstracting from their logical type thus leading a step ahead the state of the art of ontology matching.

The proposed approach has been implemented in a software that is currently being evaluated. We are evaluating the resulting alignments in a both *direct* and *indirect* way. The benchmarks used for assessing ontology matching systems are not able to evaluate the capability of finding correspondences among ontology entities with different logical types. In order to accomplish this purpose we are extending the existing benchmarks for ontology matching. On the other hand, we are using the proposed approach in a question answering system for selecting relevant resources answering a given question. The frame occurrences in a question together with the frame-ontology alignment help in formulate the query over the linked data, hence identifying resources that answer the given question.

References

1. Agirre, E., Soroa, A.: Personalizing pagerank for word sense disambiguation, pp. 33–41. ACL (2009)
2. Fillmore, C.J.: Frame semantics, pp. 111–137. Hanshin Publishing Co. (1982)
3. Gangemi, A., Alam, M., Asprino, L., Presutti, V., Recupero, D.R.: Framester: a wide coverage linguistic linked data hub, pp. 239–254 (2016)
4. Gangemi, A., Presutti, V.: Towards a pattern science for the semantic web. Semant. Web 1(1–2), 61–68 (2010)
5. Moro, A., Raganato, A., Navigli, R.: Entity linking meets word sense disambiguation: a unified approach. Trans. Assoc. Comput. Linguist. 2, 231–244 (2014)
6. Pilehvar, M.T., Jurgens, D., Navigli, R.: Align, disambiguate and walk: a unified approach for measuring semantic similarity, pp. 1341–1351. Association for Computational Linguistics (2013)
7. Ritze, D., Meilicke, C., Šváb-Zamazal, O., Stuckenschmidt, H.: A pattern-based ontology matching approach for detecting complex correspondences, pp. 25–36. CEUR-WS.org (2009)
8. Rouces, J., Melo, G., Hose, K.: FrameBase: representing N-Ary relations using semantic frames. In: Gandon, F., Sabou, M., Sack, H., d'Amato, C., Cudré-Mauroux, P., Zimmermann, A. (eds.) ESWC 2015. LNCS, vol. 9088, pp. 505–521. Springer, Cham (2015). doi:10.1007/978-3-319-18818-8_31
9. Shvaiko, P., Euzenat, J.: Ontology matching: state of the art and future challenges. IEEE Trans. Knowl. Data Eng. 25(1), 158–176 (2013)

IKEYS: Interactive KEYword Search Dedicated to Corporate Data

Khadim Dramé[1], Grégory Smits[2(✉)], and Olivier Pivert[2]

[1] Université Assane Seck de Ziguinchor, Ziguinchor, Senegal
khadim.drame@univ-zig.sn
[2] IRISA - Université de Rennes 1, Rennes, France
{gregory.smits,olivier.pivert}@irisa.fr

Abstract. IKEYS is an interactive and cooperative system aimed to query corporate linked data that allows users define explicit and unambiguous queries. Users first express their information needs through coarse keyword queries ('track J. Morrison 1971') that may then be refined with explicit projection and selection statements involving comparison operators and aggregation functions ('title of track composed by J. Morrison before 1971'). This demonstration shows how intuitive and efficient IKEYS is to find the exact answer to enhanced keyword queries.

Keywords: Keyword search · Corporate linked data · User interaction

1 Introduction

Much work has been done on keyword search over linked data, but when querying corporate knowledge bases experts expect particular functionalities, especially related to the expressivity of the querying system. Corporate linked data generally embed more numerical data than online knowledge bases and domain experts have particular expectations regarding query functionalities. Decisions domain experts have to make on top of their corporate data are generally based on results of aggregation functions or comparison operators. Existing keyword search approaches lack of expressivity and do not allow for the definition of complex and unambiguous queries containing statements involving aggregate functions and comparison operators.

To provide more expressivity and a higher precision, IKEYS relies on user interactions to both determine the exact meaning of the user's intent and offer query refinement strategies. Moreover keyword queries being inherently ambiguous, a Constrained Natural Language (CNL) is used to explain the different candidate interpretations.

The querying process of the IKEYS system is composed of three steps: (i) experts first build coarse keyword queries describing the main concepts involved in their query (Sect. 2), (ii) candidate interpretations of the coarse query wrt. the *rdf schema* are translated using a CNL and suggested to the expert (Sect. 3), and finally (iii) refinement strategies of the projection and selection statements

© Springer International Publishing AG 2017
P. Ciancarini et al. (Eds.): EKAW 2016 Satellite Events, LNAI 10180, pp. 105–108, 2017.
DOI: 10.1007/978-3-319-58694-6_10

involved in the query are proposed to introduce aggregate functions and comparison operators (Sect. 4).

The demonstration scenario (Sect. 5) aims at showing how intuitive and efficient IKEYS is and that it constitutes a pragmatic solution to corporate linked data querying.

2   Coarse Query Definition

The query definition process starts with the definition of a coarse keyword query over a knowledge base $KB = \{C \times P \times C \cup L\}$, C being a set of resources, P a set of properties (Object Properties OP and Datatype Properties DP) and L a set of literals. A coarse query is a rough expression of an information need expressed by means of keywords corresponding to the labels attached to classes C, object properties OP and literals from KB. It is assumed that the first keyword of a coarse query indicates the nature of the information to display (the projection part of the query), whereas the other keywords are related to constraints (the selection part of the query) applied to the projected information.

As shown in Fig. 1, the field in which the coarse query is defined can only be filled in with KWs suggested by an autocompletion module. For the first keyword, the one specifying the information to display, only labels attached to classes are suggested, whereas for the rest of the query, keywords matching the beginning of a word typed by the user are suggested, ordered according to their distance wrt. previously selected keywords. A keyword may match more than one element of KB, in this case, the user is asked to solve such cases of lexical ambiguity by selecting the right interpretation as shown in Fig. 3 (left).

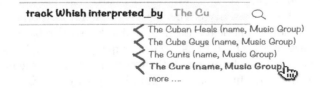

Fig. 1. Coarse query definition by autocompletion

3   Query Interpretation

The next step of the process is to determine the meaning of the coarse query by linking the elements from KB matching the keywords. The candidate interpretations of the initial coarse query correspond to all the Steiner trees that may be built on KB, rooted by the class of the projected statement, and involving an element for each of the remaining keywords. Steiner trees being not easily interpretable by a final user, syntactic rules forming a CNL are used to translate them into explicit linguistic statements as done in [4] to explain SQL queries. As shown in Fig. 2, the candidate linguistic transcriptions of the Steiner trees are suggested, in an increasing order of their size (number of edges), to the user so as to let him/her choose the right meaning.

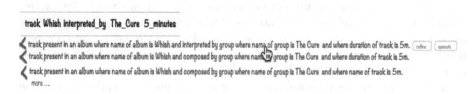

Fig. 2. User-selection of the right interpretation

4   Query Refinement

As illustrated in Fig. 3, once the right interpretation selected, the projection statement may be refined so as to precise the datatype property(ies) to display or apply aggregate functions on the class subject (e.g. the *count* function labelled *number_of*) or on a particular datatype property (e.g. the function *sum*, *max*, *average*, etc.). Selection statements may be refined as well by introducing aggregate functions and comparison operators. The refined query or more precisely its refined Steiner tree is then translated into a SPARQL query used to retrieve the answers fitting the expert's intent.

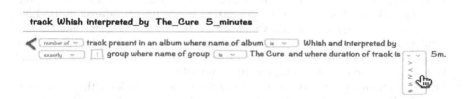

Fig. 3. Interactive query refinement

5   Demonstration Scenario

IKEYS is a Java software that may be connected to any KB whose searchable elements are linguistically labeled. IKEYS[1] acts as a proof of concept for the coarse-to-fine querying approach proposed in [3], and the demonstration scenario will allow us to show that it constitutes a pragmatic querying solution for corporate KBs that also outperforms existing approaches in terms of expressiveness and efficiency.

---

[1] Demo video at http://people.irisa.fr/Gregory.Smits/iKeys/.

For the sake of clarity, IKEYS functionalities will be demonstrated on top of the well known KB MusicBrainz that contains more than 19 millions of statements. Two *corpora* of example queries have been prepared for the demonstration scenario. The first one contains NL queries coming from the QALD-3 evaluation *corpus* [1], that will be used to show the expressivity of the approach. Starting with initial coarse queries, we will demonstrate that, thanks to the interactive disambiguation and refinement strategies of IKEYS one may express 98% of the QALD-3? *corpus*. The efficiency and relevance of IKEYS will be put into perspective wrt. the systems that have best performed during the QALD-3 challenge namely FREyA [2] and SWIP [5].

The second corpus involves more complex queries with multiple selection conditions, aggregate functions and comparison operators. These queries will be used to show the expressivity of IKEYS as well as its capability to retrieve in an efficient way the exact answer even to complex queries. As an example, we will show that, using IKEYS, it is possible to interactively build a query looking for 'the number of tracks in the album named Wish from the Cure lasting more than 5 min' and to retrieve in less than 1 second its exact answer. This kind of query cannot be expressed using classical keyword approaches whereas NL-based approaches take much more time to interpret such queries due to the inherent ambiguity of natural languages.

References

1. Cimiano, P., Lopez, V., Unger, C., Cabrio, E., Ngonga Ngomo, A.-C., Walter, S.: Multilingual Question Answering over Linked Data (QALD-3): lab overview. In: Forner, P., Müller, H., Paredes, R., Rosso, P., Stein, B. (eds.) CLEF 2013. LNCS, vol. 8138, pp. 321–332. Springer, Heidelberg (2013). doi:10.1007/978-3-642-40802-1_30
2. Damljanovic, D., Agatonovic, M., Cunningham, H.: Natural language interfaces to ontologies: combining syntactic analysis and ontology-based lookup through the user interaction. In: Proceedings of the 7th International Conference on The Semantic Web: Research and Applications, pp. 106–120 (2010)
3. Dramé, K., Smits, G., Pivert, O.: Coarse-to-fine keyword queries with user interactions. In: Proceedings of the 17th International Conference on Information Integration and Web-based Applications and Services (2015)
4. Koutrika, G., Simitsis, A., Ioannidis, Y.E.: Explaining structured queries in natural language. In: 2010 IEEE 26th International Conference on Data Engineering (ICDE), pp. 333–344. IEEE (2010)
5. Pradel, C., Peyet, G., Haemmerlé, O., Hernandez, N.: SWIP at QALD-3: results, criticisms and lesson learned. In: Forner, P., Navigli, R., Tufis, D., Ferro, N. (eds.) Working Notes for CLEF 2013 Conference, vol. 1179 (2013)

Soundness and Ontology-Based Consistency of Sensor Data Acquisition Plans

Luca Ferrari[✉], Marco Mesiti[✉], and Stefano Valtolina[✉]

Department of Computer Science "Giovanni Degli Antoni",
University of Milano, Milan, Italy
{lferrari,mesiti,valtolina}@di.unimi.it

Abstract. The verification of the soundness and consistency of data acquisition plans is an important requirement in the loading of data generated by physical and social devices. In this paper we discuss these properties in the context of the StreamLoader system.

Keywords: Internet of Things · Data streaming techniques · Distributed data integration and semantic enrichment · IoT platforms for big data

1 Introduction

Current workflow management systems (WfMS) provide users with various facilities for simplifying the development of complex programs by means of graphical interfaces. Commercial systems such as Talend Studio (www.talend.com), StreamBase Studio (www.streambase.com), Waylay.io (www.waylay.io) are designed for offering programming assistance in the design of workflows/dataflows as graphs of connected nodes representing tasks and data sources. These environments support the full application lifecycle, spanning feed integration, application modeling, development, data streams recording, testing, and debugging.

According to this design strategy, we are working on a web-based system, named StreamLoader [5], that offers facilities for the development of data acquisition plans specifically tailored for heterogeneous sensor data through the definition of a graph of services that load, filter, transform, aggregate, and compose different kinds of stored and stream data. Unlike the WfMS, StreamLoader is designed for guaranteeing the specification of dataflows that can be soundly execute, and that the generated data are semantically consistent w.r.t. a *Domain Ontology* (*DO*). A *DO* is used for describing the capabilities and properties of sensors, the act of sensing and the resulting observations in a specific domain (e.g. the analysis of meteorological conditions). In this type of analysis, the ontology should represent sensor types, location, acquisition time, and thematic definitions (e.g. the type of acquired data and its accuracy, precision, and measurement range). To do it, the SSN Ontology [3,4] has been extended with concepts/relationships that are usually adopted in the specific domain and the obtained DO can be used as constraints in the formulation of data acquisition

ⓒ Springer International Publishing AG 2017
P. Ciancarini et al. (Eds.): EKAW 2016 Satellite Events, LNAI 10180, pp. 109–113, 2017.
DOI: 10.1007/978-3-319-58694-6_11

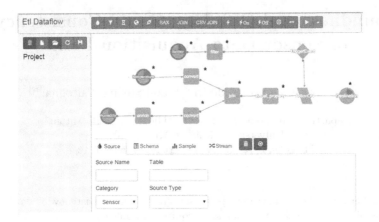

Fig. 1. Main screen of the StreamLoader web application

plans. StreamLoader adopts a very flexible, multi-granular, spatio-temporal-thematic (*STT*) data model that allows to consider heterogeneous streams of events (and stored data) generated by different kinds of sensors.

In this paper we focus on describing the set of services that can be applied for defining the dataflow and how their combination can be considered sound (w.r.t. the STT data model) and consistent (w.r.t. the adopted DO). These concepts are fundamental to guarantee that dataflows can be computed without errors, and the produced data complies with the DO along its STT dimensions.

2   Data Plan Specification

The management of complex events has been widely discussed in the past [1]. From our point of view, an event is a record of an observed change of state in the monitored situation at a given point in time. Each atomic event is characterized by the information about when it happened (time dimension), where it happened (space dimension), and what it concerned (thematic dimension). Starting from these atomic events, complex events can be generated that point out complex correlations among the basic events. To compose basic events, a set of services can be applied. Services are based on operations for the application of filters, transformation, aggregation, and composition. Operations can be classified in non-blocking (filter, cull-time/space, enrich, virtual property, transform) and blocking (aggregation, union, join, trigger on/off, convert). The formers are directly applied on each tuple when they are processed, whereas the others are processed according to time-based windows. In applying these services, we consider spatial and temporal types at different granularities that can be exploited for the specification of sound plans (further details in [5]).

Through the visual interface in Fig. 1, StreamLoader allows users to drag and drop icons representing sensors on a canvas and connect them by using the operations made available for the specification of the dataflow. In the depicted situation, when the average number of tweets in the last hour about hot temperature is greater than 20, the apparent temperature is calculated by considering temperatures and humidities identified in the city zones. Since the corresponding devises gather events at different temporal granularities, the need arises to convert them at hour granularity. Moreover, the enrich operation allows to add spatio-temporal information of where the humidity information is acquired.

3   Sound and Consistent Specification

Each service used in a dataflow is characterized by several constraints that are exploited for the verification of the service applicability that take into account the use of the spatio-temporal granularities. According to this consideration a data plan is *sound* if for each applied service the number of input streams is equal to the number of expected input streams; the parameters are specified; and, the applicability conditions on the input stream are verified.

The DO is generated by the domain experts by specifying a precise meaning of the spatio-temporal-thematic dimensions that are used in their context. This means that the structural part of the Ontology contains the spatial/temporal granularities that are adopted, the thematics that are recognized along with their attributes and types. Moreover, instances are included for representing what things should be looked up to support the creation of meaningful streams that is, sensors, observations, and related concepts.

A sound data plan is *consistent* when the schema of the output stream of generated events is consistent w.r.t. the DO. Therefore, in our environment is tolerated that the input streams or internal operations generate streams that are not consistent w.r.t. the DO, but the final stream should be consistent. This definition allows to face the issue that some devices may not produce events according to the STT dimensions, but the operations contained in the dataflow can transform the stream in one whose semantics is well-described in the DO. For examples the temperature sensors that are disseminated in a zone of the city can produce simply the observed values with no information about the time and position of the sensors. However, the gateway in charge of acquiring their observations can calculate the average temperature once a hour and assigns its position as location (at the **zone** spatial granularity) and the current hour as temporal dimension (at the **hour** temporal granularity).

Each time a sensor is virtualized in our environment, its data schema is mapped to the concepts of the DO along with the STT dimensions. If all dimensions are specified and properly mapped on the Ontology concepts, we can consider the produced event stream consistent. However, we also accept data sources for which the consistency is not verified. Indeed, other operations can be applied on them to make them consistent. Whenever the user needs to create a data acquisition plan, she introduces in the canvas icons representing the sensors and the services for their manipulation. Whenever the graph corresponding to the current dataflow is sound according to our definition, the consistency is checked starting from the data sources and moving toward the final node that collects the output data stream. For each node of the graph, new instances can be introduced in the DO (when required) in order to maintain their description at the Ontological level. Specifically, the services filter, cull-time/space do not modify the data model and then the Ontology Instances are left unchanged. By contrast, for the other services (transform, enrich, virtualProperty, aggregation, union, join, convert and trigger) it is necessary to apply a set of instructions for modifying the Ontology Instances according to the service specification. For their proper handling, a virtual sensor is introduced whose properties depend on the applied operator and on the input streams. The virtual sensor is obtained in two steps. First, the incoming sensor is cloned and renamed. Then, the cloned instance and its links are modified to comply with the operator specification.

If the initial dataflow is consistent, since each service adds or clones existing instances or associations, these services do not jeopardize the consistency with respect to the DO. This is guaranteed by the fact that, if one of the services of enrich, virtualProperty and transform creates an instance of a class not foreseen in the DO, the service fails to modify the Ontology. In the other cases, the application of the services aggregation, join, trigger, union, convert does not affect the consistency of the Ontology because they only add new associations or instances of existing classes. However, if the initial consistency of the sources is not verified, it can become consistent after the application of other operations. In Fig. 1, a * denotes that a service is consistent w.r.t. the DO. The dataflow is consistent even if some of the nodes of the graph are not consistent.

4   Conclusions

In this paper we discussed the concepts of sound and consistent specification of data acquisition plans for streams of sensor data. Once the dataflow is specified it is executed on a distributed system during the transfer of the data from the sources to the destination [2]. In the poster we will provide an overview of the StreamLoader architecture and the adopted data model. Then, we will present the adopted services with their meaning and constraints. Finally we will discuss examples of data acquisition plans that are only sound or sound and consistent with respect to a given DO in the context of meteorology. Future works aim at studying ontology-based approaches supporting stream reasoning. The idea is to help domain experts in detecting possible operators to apply in StreamLoader for enriching inconsistent sensor data specifications. In this way we can use our *DO* for providing explicit semantics for events but at the same time, for reasoning on the correctness of semantic integrations and extensions.

References

1. Cugola, G., Margara, A.: Processing flows of information: from data stream to complex event processing. ACM Comput. Surv. **44**(3), 62 pages (2012). Article No. 15
2. Zettsu, K.: Service-controlled networking. J. Nat. Inst. Inf. Commun. Technol. **62**(2), 177–184 (2015)
3. W3C Semantic Sensor Network Group: Semantic Sensor Network Ontology (2005)
4. Compton, M., et al.: The SSN ontology of the W3C semantic sensor network incubator group. Web Semant. Sci. Serv. Agents World Wide Web **17**, 25–32 (2012)
5. Mesiti, M., et al.: StreamLoader: an event-driven ETL system for the on-line processing of heterogeneous sensor data. In: Proceedings of International Conference on Extending Database Technology, pp. 628–631 (2016)

Selecting and Tailoring Ontologies with JOYCE

Erik Faessler[1]([⊠]), Friederike Klan[2], Alsayed Algergawy[2], Birgitta König-Ries[2], and Udo Hahn[1]

[1] Jena University Language & Information Engineering (JULIE) Lab,
Friedrich-Schiller-Universität Jena, Jena, Germany
{erik.faessler,udo.hahn}@uni-jena.de
[2] Heinz-Nixdorf Endowed Chair for Distributed Information Systems,
Friedrich-Schiller-Universität Jena, Jena, Germany
{friederike.klan,alsayed.algergawy,birgitta.koenig-ries}@uni-jena.de

Abstract. We present JOYCE, a scalable tool for identifying and assembling relevant (pieces of) ontologies from a repository of source ontologies, thus enabling the effective and efficient reuse of formalized domain knowledge. JOYCE includes a conceptual filter to identify relevant classes, minimizes unintended redundancies, i.e. concept duplicates, and excludes knowledge considered irrelevant for the specific conceptual design task.

Keywords: Ontology selection · Ontology modularization · Ontology reuse

1 Introduction

The rise of the Semantic Web has spurred the development of a plethora of ontologies for diverse application domains. This is particularly evident in the life sciences where portals such as BIOPORTAL serve as repositories for hundreds of domain-specific ontologies. In the light of the considerable efforts invested in the creation of these pieces of knowledge, it is desirable that these precious resoures are not just shared, but actually reused in whole or parts as a knowledge base for future applications. However, identifying relevant ontologies for a specific application and adapting them to ones own needs is a challenging task requiring a tremendous amount of time, intellectual effort and expertise. In this demo, we showcase the **J**ena **O**ntolog**Y** **C**ustomization **E**ngine (JOYCE), a scalable tool that alleviates these problems by supporting knowledge engineers in selecting and tailoring ontologies for reuse. It improves on the ONTOLOGY RECOMMENDER provided on BIOPORTAL by enabling the recommendation of ontology modules, by minimizing irrelevant and duplicate knowledge and by being more scalable in terms of the number of ontologies that can be combined.

2 Tool Overview

Input to the system can either be a set of keywords deemed relevant by domain experts or indicative paragraphs extracted from relevant scientific publications

---

E. Faessler and F. Klan—Joint First Authors.

P. Ciancarini et al. (Eds.): EKAW 2016 Satellite Events, LNAI 10180, pp. 114–118, 2017.
DOI: 10.1007/978-3-319-58694-6_12

(which undergo shallow linguistic analysis). Based on such an interest profile, JOYCE matches keywords with concept names to identify relevant (portions of) ontologies and tries to determine optimal combinations of these. It thereby aims at (a) maximizing the number of covered keywords and at minimizing (b) irrelevant and (c) duplicate knowledge. Objective (a) ensures that JOYCE assembles sets of subontologies that jointly form a reasonable basis for a knowledge base fitting the interest profile, (b) avoids superfluous representation structures in the resulting knowledge base, and (c) decreases the risk of logical inconsistencies due to incompatible conceptual models of the same concept. The latter condition ensures that the effort for integrating the ontologies or parts of them into a logically consistent knowledge base (not yet supported) by aligning semantically equivalent concepts is kept as low as possible. The importance of the criteria (a), (b) and (c) can be adjusted.

Ontology Repository and Preprocessing. We use BIOPORTAL as our base repository since it offers a wide range of (519, as of October 2016) ontologies. In order to create a local repository for a particular build of the tool we download all ontologies from BIOPORTAL. Since BIOPORTAL also hosts private ontologies, not available for public download, and conversion or parsing problems occurred, the maximum number of ontologies decreases substantially. In total, our local repository currently contains 323 ontologies. From these ontologies, we generate a dictionary of concept names for all ontology classes. This dictionary serves as an index of class names and all their synonyms as they appear in the ontologies. At the beginning of the selection process, the user-provided terms are matched with this index to determine matching candidate ontologies for the subsequent assembly process. Ontologies with an input coverage of zero are discarded.

Ontology Modularization. Ontology modularization addresses the problem of identifying portions of an ontology. The identification of a single portion given the profile terms is called *module extraction*, whereas the process that modularizes the ontology into a set of partitions is called *ontology partitioning*. While the former finds a subset of an ontology that is relevant to the terms from the interest profile, the latter reflects ontological connectivity criteria. JOYCE is able to handle both types of modules. In its current version, it incorporates a logic-based module extractor [2] (locality-based modules) and a partitioning-based modularizer [1] (partitioning-based modules). Since the latter do not depend on the interest profile, partitioning is done once when bootstrapping the system. The resulting modules are stored in a repository similar to the original ontologies. If extraction-based modules shall be combined, the tool first identifies relevant ontologies (without modifications) and then extracts locality-based modules from these relevant ontologies for subsequent assembly.

Optimization. During the assembly of candidate ontologies or ontology modules, JOYCE uses the following three optimization criteria to measure the fitness of a certain combination of ontologies or ontology modules: the proportion of irrelevant concepts (*overhead*)[1], the fraction of concepts appearing in two or more

---

[1] Modules shall comprise relevant concepts only, but often include irrelevant classes.

ontologies (*overlap*) and *term coverage*, measuring the fraction of input terms for which a matching concept was found, all relative to the number of input terms. These criteria comply with the objectives (a)–(c) discussed above. The overall goal is to maximize coverage and minimize overlap and overhead.

Identifying Optimal Combinations. We determine sets of ontology modules that are *pareto-optimal* w.r.t. the criteria coverage, overhead, and overlap. This means that they are not dominated by other solutions that are better with respect to at least one criterion and equally well with regard to the others. We also take preferences regarding the importance of these criteria into account. A knowledge engineer might, e.g., indicate that coverage and overlap are equally important. Yet, both are less important than overhead. Since computing and evaluating all possible combinations of ontology modules is infeasible (the problem is a variant of the well-known set cover problem), we use a greedy algorithm to assemble promising sets in an incremental way. We start with singleton sets of ontology modules (one for each candidate module) and add another ontology module from the candidate modules during each round. In each step, we remove those sets that are dominated by others and focus on those that fit the given preferences. To avoid the generation of a huge number of solutions, we consider just a random sample of the ontology modules that might potentially be combined with the candidate module sets determined in the previous step. The sampling process accounts for the potential of a candidate module to maximally improve the candidate sets with respect to the optimization criteria when added and also accounts for the chosen preferences. The sample size can be configured. We consider a newly discovered combination, if it is better w.r.t. at least one criterion (since otherwise it is dominated by the original set), and if its coverage increases compared to the original set. Newly encountered combinations are added to the pool of candidate sets which are input to the next round. The algorithm stops, if no new combinations are encountered or a given maximum number of iterations is reached. The same procedure works for entire ontologies.

3   User Interface and Demonstration

JOYCE comes with a Web interface based on APACHE TAPESTRY 5, which is divided into two parts: a configuration section (Fig. 1 upper part) and a results section (Fig. 1 lower part). In the configuration section, the user enters a set of keywords or a text fragment that describes the area of interest. The maximum number and the type of objects to be combined (either entire ontologies, extraction-based modules or partition-based modules), the size of the random sample taken in each assembly round, and the importance of each of the optimization criteria, *coverage*, *overlap* and *overhead*, can be adjusted appropriately. Preference values can be *low*, *medium* or *high*. The results view provides a list of the ontology (module) combinations suggested by JOYCE, sorted by coverage. For each combination, a list of the assembled ontologies or modules, their number, as well as specific values for coverage, overhead and overlap for each

Fig. 1. JOYCE – Graphical user interface

combination are displayed, thus informing the knowledge engineer about the impact of the criteria applied for tailoring. In the demo (available online on the JOYCE Web page), we will briefly introduce the user interface and the configuration parameters of JOYCE. The system will be run with various inputs and different parameter configurations to demonstrate their effects on the suggested combinations. In particular, we will change the types of objects to be combined to reveal the implications for overhead. We will gradually increase the number of objects to be assembled and look how this affects coverage. We will showcase that JOYCE can even deal with a large number of input terms and objects. Finally, we play with different preference settings for coverage, overlap and overhead, and investigate how these parameters influence the suggested combinations.

Acknowledgments. The work has been partly funded by the *Deutsche Forschungsgemeinschaft (DFG)* as part of the CRC 1076 AQUADIVA.

References

1. Algergawy, A., Babalou, S., Klan, F., König-Ries, B.: OAPT: a tool for ontology analysis and partitioning. In: International Conference on Extending Database Technology, pp. 644–647 (2016)
2. Grau, B.C., Horrocks, I., Kazakov, Y., Sattler, U.: Just the right amount: extracting modules from ontologies. In: International World Wide Web Conference, pp. 717–726 (2007)

Representing Contextual Information as Fluents

José M. Giménez-García[1]([✉]), Antoine Zimmermann[2], and Pierre Maret[1]

[1] Univ Lyon, UJM-Saint-Etienne,
CNRS, Laboratoire Hubert-Curien UMR 5516, 42023 Saint-Étienne, France
`jose.gimenez.garcia@univ-st-etienne.fr`
[2] École Nationale Supérieure des Mines, FAYOL-ENSMSE,
Laboratoire Hubert Curien, 42023 Saint-Étienne, France

Abstract. Annotating semantic data with metadata is becoming more and more important to provide information about the statements. While there are solutions to represent temporal information about a statement, a general annotation framework which allows representing more contextual information is needed. In this paper, we extend the 4dFluents ontology by Welty and Fikes to any dimension of context.

1   Introduction

In the Semantic Web, it is often necessary to characterize the context associated to a statement, *e.g.*, when it was generated, or who said it. In RDF and OWL this can only be represented natively using binary relations. There are generic approaches for representing statements about statements, such as using reification[1], N-ary relations[2] or N-Quads[3], and the Singleton Property [4]. However, each of them has its own drawbacks. When using reification, inference is prevented; in the case of N-ary relations, the structure of the domain ontology needs to be changed (not always possible when reusing external ontologies). For N-Quads there are no formal semantics, and its usage for named graphs has been standardized. The Singleton Property on the other hand, requires to extend the formal semantics of RDF and, due to the explosion in the number of properties, has proved not to be efficient in current knowledge bases [3].

On the other hand, Welty and Fikes [6] propose a model to represent temporal validity of an entity by considering it a perdurant, without any of the previous disadvantages. This work has been used by the community and further extended by other authors to address the proliferation of slices [7], or representing spatiotemporal information [1]. While Welty [5] proposes a generalization of the model by substituting the temporal part of an entity by a contextual projection, he does not address the possibility of using different dimensions of context in the same dataset.

In this work, we propose an extension of Welty and Fikes model to a generic ontology that can be extended to implement any number of concrete dimensions

---

[1] http://www.w3.org/TR/2014/REC-rdf11-mt-20140225/#reification.
[2] https://www.w3.org/TR/swbp-n-aryRelations.
[3] https://www.w3.org/TR/n-quads.

© Springer International Publishing AG 2017
P. Ciancarini et al. (Eds.): EKAW 2016 Satellite Events, LNAI 10180, pp. 119–122, 2017.
DOI: 10.1007/978-3-319-58694-6_13

of context (Sect. 2). In addition, we address different issues and decisions to make when modeling a knowledge base using out approach (Sect. 3). Finally, we give some final remarks and possible lines of future work (Sect. 4).

2   NdFluents

Welty and Fikes [6] address the problem of representing *fluents, i.e.*, relations that hold only within a certain time. They address the issue by using the perdurantist view. According to it, entities are four dimensional constructs, and instead of making statements about them, one should make the assertions about their temporal parts. Instead of making an assertion about some entities, such as *"Paris is the capital of France"*, one should make the assertion about their temporal parts: *"A temporal part of Paris (since 508 up to now) is the capital of a temporal part of France (since 508 up to now)"*.

The temporal part of an entity can be viewed as an individual context dimension of the entity. A similar approach can then be used to represent different dimensions, such as provenance. Continuing with our running example, we can articulate that fact as *"Paris (according to Wikipedia) is the capital of France (according to Wikipedia)"*. Different context dimensions of an entity could then be combined, allowing to represent complex information: *"According to Wikipedia, Paris has been the capital of France since 508"*.

We use this idea to generalize the 4dFluents ontology for any context dimension in the *NdFluents* ontology [2]. The ontology, shown down below, is a direct extension from temporal parts to any contextual parts. While approaches that reify the predicate hinder OWL reasoning, NdFluents allows for OWL inference of OWL property axioms within the same contexts. The ontology, the extensions for temporal and provenance dimensions, and a use case where both dimensions are used (the estimated evolution of Earth population according to different sources) is published in http://www.emse.fr/~zimmermann/ndfluents.html

```
1    Declaration( Class( nd:Context ) )
2    Declaration( Class( nd:ContextualPart ) )
3    DisjointClasses( nd:Context nd:ContextualPart )
4    Declaration( ObjectProperty( nd:contextualProperty ) )
5    ObjectPropertyDomain( nd:contextualProperty nd:ContextualPart )
6    ObjectPropertyRange( nd:contextualProperty nd:ContextualPart )
7    Declaration( DataProperty( nd:contextualDatatypeProperty ) )
8    DataPropertyDomain ( nd:contextDataProperty nd:ContextualPart )
9    Declaration( ObjectProperty( nd:contextualExtent ) )
10   ObjectPropertyDomain( nd:contextualExtent nd:ContextualPart )
11   ObjectPropertyRange( nd:contextualExtent nd:Context )
12   Declaration( ObjectProperty( nd:contextualPartOf ) )
13   FunctionalObjectProperty( nd:contextualPartOf )
14   ObjectPropertyDomain( nd:contextualPartOf nd:ContextualPart )
15   ObjectPropertyRange( nd:contextualPartOf ObjectComplementOf( nd:Context ))
```

3   Modeling a Knowledge Base with NdFluents

In this section we present possible issues that may arise from the approach and what can be done to solve them when modeling data using the NdFluents ontology. While the first two points are common to Welty and Fikes [6] approach, the last ones are only relevant when using more than one dimension.

Adapting Non-Contextual Ontologies to NdFluents: When reusing existing ontologies, it is not always possible to add a subproperty relationship between an already defined property and a contextual property, because most restrictions will not hold for the contextual parts of an entity. For instance, let us suppose that the property `capitalOf` has domain `City`, and we use it as a fluent property. Then, it would be inferred that `Paris@1` is a city, instead of `Paris` being inferred as a city.

Dealing with Terminological Statements: In general, contexts are used just for assertions (the *ABox*). While using contexts with terminological statements (the *TBox*) is possible, it is important to take into account that new properties will not benefit from the standard inferences associated with `subClassOf`.

Relations between Different Contextual Parts: The NdFluents ontology allows to model relations among different contextual parts of different context dimensions (*i.e.*, a temporal part of Paris could be the capital of a provenance part of France). If it is needed for a contextual property to be related to contextual parts of the same dimension of context, it will be necessary to add the appropriate axioms to the ontology:

```
1   Declaration( ObjectProperty( 4d:fluentProperty ) )
2   SubObjectPropertyOf( 4d:fluentProperty nd:contextualProperty )
3   ObjectPropertyDomain( 4d:fluentProperty 4d:TemporalPart )
4   ObjectPropertyRange( 4d:fluentProperty 4d:TemporalPart )
5   Declaration( DataProperty( 4d:fluentDataTypeProperty ) )
6   SubDataPropertyOf( 4d:fluentDataTypeProperty nd:contextualProperty )
7   DataPropertyDomain( 4d:fluentProperty 4d:TemporalPart)
```

Combining Different Context Dimensions: An important scenario where NdFluents becomes relevant is when the necessity of combining two or more dimensions of context arises. There are different possibilities to model them in NdFluents.

– *Contexts in Context:* Relating a contextual part to another contextual part. This approach can be taken when the "first level" contextual parts are relevant facts of the knowledge base, and we want to state additional information about them. To improve reasoning, `contextualPartOf` property needs to be transitive, which can be achieved by adding the following axiom:
  `TransitiveObjectProperty( nd:contextualPartOf )`
– *Use Multiple Contexts for each Contextual Part:* Only one contextual part is created for a combination of context dimensions. This contextual part is then related to all the related contextual information. This model is easier to model: Relating the contextual part with the context dimensions is straightforward. It also avoids ambiguity when modeling contextual information related to more than one contextual dimension, and reduces the number of resources in the ontology (*i.e.*, while the previous model needed one contextual part for each context dimension involved, this approach only requires one contextual part).

Note that `contextualPartOf` is a functional property, which means that there cannot be a contextual part of more than one entity.

- *Relations between Different Contextual Parts:* Creating contextual extents that combine two or more context dimensions, and enforcing a limit of only one contextual extent per contextual part. This model adds a layer of complexity to the previous approach, but it can be useful to require a specific combination of context dimensions on a set of contextual parts. This can be achieved by adding the axiom `FunctionalObjectProperty(` nd:contextualExtent `)`

4   Conclusion

Representing contextual information in different dimensions is a current challenge in OWL. We have proposed NdFluents, a multi-domain contextual representation, generalizing the 4dFluents ontology to any number of context dimensions. NdFluents allows for a more complete OWL inference than other generic approaches, and allows to retrieve easily all the information within a context for the same entity.

Acknowledgments. This project is supported by funding received from the European Union Horizon 2020 research and innovation program under the Marie Skłodowska-Curie grant agreement No 642795.

References

1. Batsakis, S., Petrakis, E.G.M.: Representing temporal knowledge in the semantic web: the extended 4D fluents approach. In: Hatzilygeroudis, I., Prentzas, J. (eds.) Combinations of Intelligent Methods and Applications. SIST, vol. 8, pp. 55–69. Springer, Heidelberg (2011)
2. Giménez-García, J.M., Zimmermann, A., Maret, P.: NdFluents: A Multidimensional Contexts Ontology. Univ. Jean Monnet, Technical report (2016)
3. Hernández, D., Hogan, A., Krötzsch, M.: Reifying RDF: what works well with wikidata? In: Proceedings of the 11th International Workshop on Scalable Semantic Web Knowledge Base Systems, pp. 32–47 (2015)
4. Nguyen, V., Bodenreider, O., Sheth, A.P.: Don't like RDF reification?: making statements about statements using singleton property. In: WWW, pp. 759–770. ACM Press (2014)
5. Welty, C.: Context slices: representing contexts in OWL. In: Workshop on Ontology Patterns. CEUR Workshop Proceedings, vol. 671, p. 59. CEUR-WS.org (2010)
6. Welty, C., Fikes, R.: A reusable ontology for fluents in OWL. In: Proceedings of the First International Conference of Formal Ontology in Information Systems (FOIS). Frontiers in Artificial Intelligence and Applications, vol. 150, pp. 226–236. IOS Press (2006)
7. Zamborlini, V., Guizzardi, G.: On the representation of temporally changing information in OWL. In: Enterprise Distributed Object Computing Conference Workshops (EDOCW), pp. 283–292. IEEE (2010)

An Ontological Perspective on Thematic Roles

Anna Goy[✉][iD], Diego Magro[iD], and Marco Rovera[iD]

Dipartimento di Informatica, Università di Torino, Torino, Italy
{annamaria.goy,diego.magro,marco.rovera}@unito.it

Abstract. We face the issue of formally modeling roles in the semantic representation of events: we propose a distinction between *thematic roles* and *social roles*, and we show that they have important ontological differences, suggesting distinct formal representations. We apply our approach in the context of the Harlock'900 project, including the definition of thematic roles as *binary properties* in the HERO ontology.

Keywords: Ontological analysis · Event ontology · Thematic roles · Ontology design patterns

1 Introduction

In the Semantic Web and ontology modeling community, the representation of *roles* has being always considered an important challenge. However, when discussing this issue within different specific fields, the notion of *role* has been interpreted in many, different, and sometimes confusing ways. In this paper we face the issue of formally modeling roles in the perspective of the ontological representation of events: we briefly mention the most relevant related work (Sect. 2), we discuss our ontological analysis, and we propose a formal representation for thematic roles (Sect. 3). We conclude by summarizing the main future directions of our work (Sect. 4).

2 Related Work and Background

In this section, we briefly sketch the modeling choices concerning roles that can be found in some well-known (event) ontologies, and an (incomplete) outline of the extremely wide literature about thematic roles in the (computational) linguistics field.

None of the analyzed ontologies provides a fine-grained account of the different ways in which entities can participate in events. However, all of them provide some means to represent general participation (e.g., EDM [1] among many others), and some of them also offer a *pattern* for representing roles that can be used to express specific modalities of event participation in conjunction with external resources (e.g., SEM [15], CIDOC-CRM [5], Event Model F [14]). Moreover, none of the analyzed ontologies make any (formal) distinction between roles played by event participants and social roles (see Sect. 3).

In the perspective taken in this paper, undoubtedly influential works in the (computational) linguistic area are those by Pustejovsky [12], Jackendoff [4], and Levin and Rappaport [6], claiming that thematic roles are not semantic primitives, but relations

© Springer International Publishing AG 2017
P. Ciancarini et al. (Eds.): EKAW 2016 Satellite Events, LNAI 10180, pp. 123–126, 2017.
DOI: 10.1007/978-3-319-58694-6_14

between individuals and events, possibly emerging from rich semantic structures. Another significant thread is *Frame Semantics*, introduced by Fillmore [2], and representing the starting point of the FrameNet project (framenet.icsi.berkeley.edu), a huge English lexical database, where word meanings correspond to *frames*, representing event types, and including participants playing different roles. Finally, an important debate concerns thematic roles specificity: are thematic roles specific for every type of event (*buyer, seller,...*) [10], or are they general ways of participating in events of different kinds (*agent, patient,...*) [3, 6]?

3   An Ontological Approach to Thematic Roles

Our analysis of the literature and ontological models led us to identify at least two different notions that have been called "roles", but show a quite different nature:

- The role somebody/something plays when (s)he/it **participates in an event** ("Brutus killed Caesar with a knife", where Brutus plays the role of *killer/agent*, Caesar that of *killed/patient*, and the knife plays the role of *instrument*) – sense (a).
- The role somebody plays within a given **social context** ("Renzi is the current Italian Prime Minister", where no event is explicitly mentioned, but there is clearly somebody who plays a role, that of Italian Prime Minister) – sense (b).

Obviously, these different senses of the notion of role are not totally independent; for example, if somebody kills someone (sense (a)), it can be socially considered a *killer* (sense (b)); or somebody can participate in an international meeting (sense (a)) "qua" Italian Prime Minister (sense (b)). However, we will show that:

1. Besides similarity, sense (a) and (b) show important **ontological differences**.
2. As a consequence, they should be modeled in different ways and roles in sense (a) can be modeled as **binary properties**, connecting the event to its participants.

Following Masolo and colleagues [9], we characterize roles as *anti-rigid* and *founded* concepts. Roles (both social and thematic ones) are *anti-rigid* [16] since any individual that plays a role does not play it *necessarily*. Anti-rigidity also implies that any entity can start and stop playing a role, or it can change role during its life: e.g., no Prime Minister *necessarily* is a Prime Minister, and any Prime Minister starts and stops being a Prime Minister at a given time in her/his life. Similarly, no patient in an event *necessarily* is a patient, (s)he/it stops being a patient when the event ends and there can be time periods during which (s)he/it is not patient in any event. Moreover, in general, an entity can play the same (social or thematic) role several times, a (social or thematic) role can be played by different entities at the same time and an entity can play different roles simultaneously. Moreover, (social) roles are *founded* concepts [9]: intuitively, a concept *x* is founded if and only if its definition mentions another concept *y*, "such that for each entity classified by *x*, there is an entity classified by *y* which is external to it"[1]

---

[1] "Externality" is actually a complex notion that, in many practical cases, can be conveniently approximated by stating that *y* is external to *x* iff *x* is not part of *y* and *y* is not part of *x*.

[9]. It is easy to see that foundedness holds also for thematic roles, since any definition of a thematic role necessarily mention the notion of event (e.g., the *beneficiary* role is typically defined as "the entity taking advantage *of an event*") and each particular event is external to any entity that can play a thematic role in it. As a consequence, both thematic and social roles are *roles* in the sense formalized by Masolo and colleagues [9], but social roles cannot be equated with the roles played when participating in an event: in fact, if somebody is a musician (social role), (s)he is still a musician also when sleeping (i.e., when (s)he is not participating in any event in which (s)he acts as a musician) [7]. The distinction between thematic and social roles can be further supported by considering the relation between an event and its participants which is bounded by the temporal boundaries of the event itself (if I participate in a hitting event as a *patient*, I stop being a *patient* when the event ends).

The ontological differences between thematic roles (sense (a)) and social roles (sense (b)) underpin our claim that they should be formalized in different ways. In computational ontologies, roles are usually *reified* and placed in the domain of discourse, in order to offer models that enable one to "talk about roles" and to explicitly represent relations they are involved in. Moreover, sometimes, also *role attributions* (i.e. the relationships between entities and the roles they play) are reified, usually in order to be able to assign temporal boundaries to the relation between an entity and the (social) role it plays (see, for instance, the Publishing Role Ontology [11]). Our hypothesis is that the formal representation of events requires a model that allows us to *use* social and thematic roles, and to *talk about* social roles, while we do not need to *talk about* thematic roles. In particular, we do not need the role attribution pattern mentioned above for thematic roles, since the events themselves already provide the spatio-temporal context required; therefore, while social roles should be conveniently reified, we do not need to reify thematic roles. We think that the nature of thematic roles is better formally accounted for by means of *binary properties*, providing a more immediate representation of the close relationships between events and participants.

In order to verify our hypothesis, we applied our approach in the specific context of Harlock'900 [3, 13], a project (2016–2018) involving the Computer Science Dept. of the University of Torino and the Fondaz. Ist. Piemontese A. Gramsci, a member of the Polo del '900 (www.polodel900.it). Within Harlock'900, we are developing HERO, an Historical Event Representation Ontology relying on DOLCE [8]. As an example, we show here the axioms defining the generic notion of participation in an event and the "classical" *patient* role (free variables are universally quantified):

```
hero:object(x) → dolce:endurant(x)
hero:event(x) → dolce:event(x)
hero:hasParticipant(x,y) → hero:event(x) ∧ hero:object(y)
hero:hasParticipant(x,y) → (∃t)(dolce:timeInterval(t) ∧
                               dolce:participatesIn(y,x,t))
hero:hasPatient(x,y) → hero:event(x) ∧ hero:object(y)
hero:hasPatient(x,y) → hero:hasParticipant(x,y)
```

4   Conclusions and Future Work

In this paper, we provided an ontological analysis of the notions of *social* and *thematic role*, arguing for the suitability of representing thematic roles as properties connecting events and their participants. Obviously, a lot of work can start from here. For example, the issues about *general* versus *specific* thematic roles should be taken into account. We are also building an OWL version of the HERO ontology, to be used to test our approach. Finally, another related notion deserves further investigation: the role attribution expressing a *point of view* ("The liberators landed at Sicily", where the Anglo-American allies are seen as liberators, from a specific historical perspective).

References

1. EDM: EDM Definition of the Europeana Data Model v.5.2.7 (2016). pro.europeana.eu
2. Fillmore, C.J.: Frame semantics. In: The Linguistic Society of Korea (ed.) Linguistics in the Morning Calm, pp. 111–137, Hanshin, Soeul (1982)
3. Goy, A., Magro, D., Rovera, M.: Ontologies and historical archives: a way to tell new storie. Appl. Ontology **10**(3–4), 331–338 (2015)
4. Jackendoff, R.: Semantic Structures. MIT Press, Cambridge (1990)
5. Le Boeuf, P., Doerr, M., Ore, C.E., Stead, S. (eds.): Definition of the CIDOC Conceptual Reference Model, v. 6.2.1. ICOM/CIDOC CRM SIG (2015)
6. Levin, B., Rappaport, M.: Wiping the slate clean: a lexical semantic exploration. In: Levin, B., Pinker, S. (eds.) Lexical and Conceptual Semantics, pp. 123–151. Blackwell, Cambridge (1991)
7. Loebe, F.: An analysis of roles. Toward ontology-based modelling. Master's Thesis, University of Leipzig (2003)
8. Masolo, C., Borgo, S., Gangemi, A., Guarino, N., Oltramari, A.: WonderWeb Deliverable D18. Technical report, CNR (2003)
9. Masolo, C., Vieu, L., Bottazzi, E., Catenacci, C., Ferrario, R., Gangemi, A., Guarino, N.: Social roles and their descriptions. In: Dubois, et al. (eds.) Proceedings of the KR 2004, pp. 267–277. AAAI Press, CA (2004)
10. McRae, K., Ferretti, T.R., Amyote, L.: Thematic roles as verb-specific concepts. Lang. Cogn. Process. **12**(2/3), 137–176 (1997)
11. Peroni, S., Shotton, D., Vitali, F.: Scholarly publishing and the linked data: describing roles, statuses, temporal and contextual extents. In: Sack, H., Pellegrini, T. (eds.) Proceedings of the i-Semantics 2012, pp. 9–16. ACM, New York (2012)
12. Pustejovsky, J.: The syntax of event structure. Cognition **41**(1–3), 47–81 (1991)
13. Rovera, M.: A knowledge-based framework for events representation and reuse from historical archives. In: Sack, H., Blomqvist, E., d'Aquin, M., Ghidini, C., Ponzetto, S.P., Lange, C. (eds.) ESWC 2016. LNCS, vol. 9678, pp. 845–852. Springer, Cham (2016). doi: 10.1007/978-3-319-34129-3_53
14. Scherp, A., Franz, T., Saathoff, C., Staab, F.: F - A model of events based on the foundational ontology DOLCE + DnS Ultralite. In: Gil, Y., Noy, N. (eds.) Proceedings of the K-CAP 2009, pp. 137–144. ACM, New York (2009)
15. van Hage, W.R., Malaisé, V., Segers, R., Hollink, L., Schreiber, G.: Design and use of the Simple Event Model (SEM). J. Web Semant. **9**, 128–136 (2011)
16. Welty, C., Guarino, N.: Supporting ontological analysis of taxonomic relationships. Data Knowl. Eng. **39**, 51–74 (2001)

StreamJess: Enabling Jess for Stream Data Reasoning and the Water Domain Case

Edmond Jajaga[1(✉)], Lule Ahmedi[2], and Figene Ahmedi[3]

[1] Department of Computer Science, South East European University, Ilindenska n. 335,
1200 Tetovë, Macedonia
e.jajaga@seeu.edu.mk

[2] Department of Computer Engineering, University of Prishtina, Kodra e diellit pn,
10000 Prishtinë, Kosova
lule.ahmedi@uni-pr.edu

[3] Department of Hydro-Technic, University of Prishtina, Kodra e diellit pn,
10000 Prishtinë, Kosova
figene.ahmedi@uni-pr.edu

Abstract. This paper introduces *StreamJess*, a Stream Reasoning system that layers on top of a state of the art query processing system such as C-SPARQL to enable closed-world, non-monotonic and time-aware reasoning with Jess rules. The system is validated in the water quality monitoring domain by demonstrating water bodies' classification and pollution sources investigation.

Keywords: Stream data · Expert system · Semantic web · Rules · Water quality monitoring · Ontologies

1 Introduction

Even though the Semantic Web technologies have been extensively used for modelling stream data domains, e.g. SSN ontology[1], and for processing through SPARQL-like extensions, e.g. C-SPARQL [1], EP-SPARQL [2], etc., the recommended rules standards, SWRL and RIF, still remain not applicable in the domain of stream data applications. As a result, the stream data knowledge bases have been merely coupled with production rules, answer set programming or event processing systems [3]. In the vision of building a unique Semantic Web platform for reasoning over stream data, we have developed a production rules system, *StreamJess*, which layers Jess rules to reason over our water quality monitoring (WQM) ontology named InWaterSense [4]. Jess supports closed-world and non-monotonic reasoning. However, extending Jess with stream data reasoning features is very expensive. Code could not be optimized even for simple temporal operations over event-streams [5]. Thus, we propose a much simpler approach by coupling stream data processing features, supported by state of the art Stream Reasoning (SR) query systems such as C-SPARQL, with Jess's reasoning abilities. C-SPARQL supports time-aware reasoning on stream data. However, as a query language,

---

[1] Semantic Sensor Network Ontology, http://purl.oclc.org/NET/ssnx/ssn.

P. Ciancarini et al. (Eds.): EKAW 2016 Satellite Events, LNAI 10180, pp. 127–130, 2017.
DOI: 10.1007/978-3-319-58694-6_15

it is not intended to have any effect on the underlying ontology. In *StreamJess* we use Jess rules for populating the knowledge base. Moreover, they enable data modifications i.e. non-monotonic reasoning and the tools for archiving data. The system is validated with simulated data in the WQM domain, but it is developed for use within the InWaterSense project[2] with real data. The simulator randomly generates observation data for an arbitrary number of 70 measurement sites and 11 water quality parameters. A single sensor observation is arbitrarily set to be produced every second and includes 6 RDF streams representing time, location, device and quality of observation information. For example, in a 20 s window size 120 tuples will be produced. Moreover, the system supports registering multiple streamers to run concurrently.

2   Conceptual Design and Implementation

As depicted in Fig. 1, *StreamJess* acts as a pipeline. Incoming RDF data streams, e.g. sensor observed values, are firstly filtered out and eventually aggregated by C-SPARQL queries. The query results are asserted into the knowledge base through JessTab functions. The ABox changes will eventually cause to fire Jess rules, which have been registered on application startup. The Jess engine inferences will be again published onto the ontology. The processing and reasoning over incoming streams is iterative for each window.

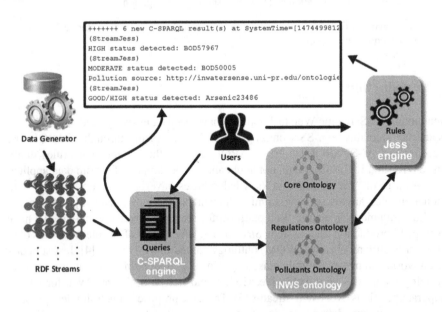

Fig. 1.   *StreamJess* architecture

---

[2]   http://inwatersense.uni-pr.edu/.

StreamJess is implemented as a Java console application. The application uses an instance of `jess.Rete` which is created at system start up. It provides the central access point of the application as it loads the ontology, builds the working memory, holds the list of rules and offers the methods for doing CRUD operations over facts [6]. Namely, Protégé functions of Jess in JessTab were used to manage with the knowledge base. All ontology modules are imported and loaded at application start up. Moreover, class instances are also mapped into the Jess's working memory. Different stream data ontologies can be loaded into the system and appropriate C-SPARQL queries and Jess rules can be defined to run over incoming data streams.

3 Validation

As a proof of concept, we have implemented *StreamJess* in a typical WQM scenario based on WSN. Sensors in InWaterSense WQM system are deployed in different measurement sites at different times. *StreamJess* will (1) classify the water body into the appropriate status (good, high or moderate) according to WFD regulations [7, 8] and (2) identify the potential sources of pollution in case of critical status detection. In general, each water quality is monitored and investigated with a *monitoring* rule (1) and an *investigation* one (2).

For brevity, we will demonstrate the case of Biochemical Oxygen Demand (BOD_5) and pH observations. Like most of water quality parameter observations, BOD_5 observations are classified based on the average value of measurements within a time interval while pH ones are considered one by one [8]. Two C-SPARQL queries are deployed into the system to match each of the types of observations. Moreover, four rules are deployed, one for monitoring and one for investigation of BOD_5 observations and another couple for pH observations. After loading all start up components, the user is asked to specify the window type of the queries. Namely, if he specifies time-based windows then he is presented with another question for setting the window size in seconds. Otherwise, he may specify to use tuple-based windows by further providing the number of tuples to be processed within a window. Each query eventually outputs triples of values: the water quality name, the location of measurements and the observed value i.e. the average value of BOD_5 measurements or individual pH measurements. Every output triple is mapped into a temporary observation class. Furthermore, for each new incoming triple a new call to the Rete method `run()` is invoked for doing rule-based reasoning. As illustrated in Fig. 1, the Jess engine runs the rules against the temporary observation facts, produced by C-SPARQL, and it eventually activates the rule's RHS actions. The inferred knowledge forms another set of RDF data which is stored back into the ontology for further reasoning. Namely, monitoring rules do the water quality classifications based on the WFD regulation rules which general form looks like follows: `{observation details} => {classify and archive the observation}`. In case a critical status is detected, investigation rules act to identify the pollution source which general form is: `{moderate status observation} => {get and display the sources of pollution present on the measurement site}`.

An output of the running example is illustrated in Fig. 2, where C-SPARQL processing of RDF streams has resulted with 3 new observations on 3 measurement sites: ms10, ms11 and ms12. Two observations have been classified as of "moderate" status (line #1 and #2) and one of "high" status (line #3). Potential sources of pollution include urban stromwater discharges and fish farming on site ms11 while urban stormwater is potential source of BOD_5 discharges on site ms10. An online demo of *StreamJess* can be found on the following link http://inwatersense.uni-pr.edu/streamjess/demo.html.

```
+++ 3 new C-SPARQL result(s) at SystemTime=[1462808878402] ++++
#1 (C-SPARQL) Observed WQ: BiochemicalOxygenDemand Value: 1.6622
(StreamJess)(42693) BOD status is MODERATE
On: Mon May 09 17:47:58 CEST 2016
In: http://inwatersense.uni-pr.edu/ontologies/inws-core.owl#ms11
BOD5 pollution source: Urban stormwater discharges
BOD5 pollution source: Fish farming
#2 (C-SPARQL) Observed WQ: BiochemicalOxygenDemand Value: 1.9215
(StreamJess)(42181) BOD status is MODERATE
On: Mon May 09 17:47:58 CEST 2016
In: http://inwatersense.uni-pr.edu/ontologies/inws-core.owl#ms10
BOD5 pollution source: Urban stormwater discharges
#3 (C-SPARQL) Observed WQ: BiochemicalOxygenDemand Value: 0.6389
(StreamJess)(50374) BOD status is HIGH
On: Mon May 09 17:47:58 CEST 2016
In: http://inwatersense.uni-pr.edu/ontologies/inws-core.owl#ms12
```

Fig. 2. An excerpt of the output of the running example

4   Conclusion

Until recently most of the SR research has been dedicated on ontology and stream processing developments. Our work goes beyond the query processing achievements and thus focusing on rule level implications of stream data reasoning. SWRL lacks the required expressivity level to reason over stream data. Thus, we built *StreamJess*, a production rule system supporting time-aware and non-monotonic reasoning.

References

1. Barbieri, D.F., Braga, D., Ceri, S., Della Valle, E., Grossniklaus, M.: C-SPARQL: a continuous query language for RDF data streams. J. Semant. Comput. **4**(1), 3–25 (2010)
2. Anicic, D., Fodor, P., Rudolph S., Stojanovic, N.: EP-SPARQL: a unified language for event processing and stream reasoning. In: WWW 2011, pp. 635–644 (2011)
3. Jajaga, E., Ahmedi, L., Abazi-Bexheti, L.: Semantic Web trends on reasoning over sensor data. In: 8th South East European Doctoral Student Conference, Greece, pp. 284–293 (2013)
4. Ahmedi, L., Jajaga, E., Ahmedi, F.: An ontology framework for water quality management. In: Corcho, Ó., Henson, C.A., Barnaghi, P.M. (eds.) SSN@ISWC, Sydney, pp. 35–50 (2013)
5. Ermert, L.: Comparing Jess and Esper for event stream processing. Bachelor Thesis, Faculty IV - department of computer science, Fachhochschule Hannover, Germany (2009)
6. Hill, E.F.: Jess in Action: Java Rule-Based Systems. Manning Publications Co., Greenwich (2003)
7. Environment Agency: Method statement for the classification of surface water bodies, v2.0 (external release) [online], Monitoring Strategy v2.0, July 2011
8. Statutory Instruments. European Communities Environmental Objectives (Surface Waters) Regulations 2015, S.I. No. 386 of 2015 (2009). http://www.irishstatutebook.ie/eli/2015/si/386/made/en/pdf

Flexible RDF Generation from RDF and Heterogeneous Data Sources with SPARQL-Generate

Maxime Lefrançois$^{(\boxtimes)}$, Antoine Zimmermann, and Noorani Bakerally

Univ. Lyon MINES Saint-Étienne, CNRS,
Laboratoire Hubert Curien UMR 5516, 42023 Saint-Étienne, France
{maxime.lefrancois,antoine.zimmermann,noorani.bakerally}@emse.fr

Abstract. RDF aims at being the universal abstract data model for structured data on the Web. While there is effort to convert data in RDF, the vast majority of data available on the Web does not conform to RDF. Indeed, exposing data in RDF, either natively or through wrappers, can be very costly. In this context, transformation or mapping languages that define generation of RDF from non-RDF data represent an efficient solution. Furthermore, the declarative aspect of these solutions makes them easy to adapt to any change in the input data model, or in the output knowledge model. This paper introduces a novel such transformation language (SPARQL-Generate), an extension of SPARQL for querying not only RDF datasets but also documents in arbitrary formats. Its implementation on top of Apache Jena currently covers use cases from related work and more, and enables to query and transform web documents in XML, JSON, CSV, HTML, CBOR, and plain text with regular expressions.

Keywords: RDF · SPARQL · Linked data · Data transformation

1 Introduction

The vision of a Semantic Web where machines can more easily process web content is hampered by the coexistence of many, heterogeneous data formats and models available on the Web or *via* the Web. At first sight, this vision seems to require a worldwide adoption of a common universal data model as a document format (namely, RDF). Yet, in the emerging Web of Data, a multitude of formats are flourishing: XML (not RDF/XML) is still very present, open data portals tend to prefer CSV, web APIs rely more on JSON, and the Web of Things introduces new formats adapted to the resource constraints inherent to less powerful devices.

This paper has been partly funded by the ITEA2 12004 SEAS (Smart Energy Aware Systems) project, the ANR 14-CE24-0029 OpenSensingCity project, and a research contract with ENGIE R&D.

© Springer International Publishing AG 2017
P. Ciancarini et al. (Eds.): EKAW 2016 Satellite Events, LNAI 10180, pp. 131–135, 2017.
DOI: 10.1007/978-3-319-58694-6_16

In this context, transformation or mapping languages that define generation of RDF from non-RDF data represent an efficient solution. Furthermore, the declarative aspect of these solutions makes them easy to adapt to any change in the input data model, or in the output knowledge model. In this paper, we use term *RDF lifting* to designate the process of generating RDF from non-RDF data.

Section 2 first overviews existing initiatives and evaluates them with respect to a set of identified requirements. Then Sect. 3 introduces SPARQL-Generate, a language that can be used to specify a mapping from distributed data sources in arbitrary data formats to the RDF data model. Section 4 concludes and identifies directions for future work.

2   Requirements for RDF Lifting Mechanisms

From the use cases faced by industrial partners in collaborative projects we are involved in, we identified the following set of requirements for a RDF lifting mechanisms. It should:

- be able to generate RDF from multiple sources in heterogeneous formats;
- be able to deal with text-based *and* binary representation formats;
- make it easy to combine non-RDF sources with RDF data;
- be extensible to account for new syntaxes;
- integrate seamlessly with existing standards for consuming Semantic Web data, such as SPARQL or Semantic Web programming frameworks.

Several systems exist for converting data to RDF. We are interested only in systems that provide a formal transformation language, potentially mapping-based.[1] Such languages include GRDDL [1], XSPARQL [5], R2RML [2], RML [3], and CSVW [6]. Table 1 overviews the comparison of these solutions with respect to the identified requirements. GRDDL and XSPARQL rely respectively on XSLT and XQuery, that have been proven to be Turing-complete. These languages are hence full-fledged procedural programming language with explicit algorithmic constructs to produce RDF. We argue that a procedural paradigm is less suited to semantic web engineers, who may be more familiar with a declarative paradigm such as SPARQL. Furthermore, only RML and XSPARQL are specifically dedicated to generate RDF from various formats, namely XML, CSV, HTML, and JSON. However, to our knowledge, they do not accept binary formats such as EXI, CBOR or BSON, which become of much importance in the emerging Web of Things.

3   SPARQL-Generate

SPARQL-Generate extends SPARQL 1.1 with only three new clauses, generate, source and iterator. It queries the combination of an RDF dataset and zero or

---

[1] A lot of hardcoded transformation are available for many formats – https://www.w3.org/wiki/ConverterToRdf.

Table 1. Features of related work compared to the SPARQL Generate language.

	GRDDL [1]	CSVW [6]	R2RML [2]	XSPARQL [5]	RML [3]	SPARQL-Generate
Produces RDF	✓	✓	✓	✓	✓	✓
Multiple source		✓	✓	✓	✓	✓
Heterogeneous formats				✓	✓	✓
Binary formats						✓
Combines RDF data				✓	✓	✓
Extensibility	XSLT			XQuery	ad-hoc	SPARQL

more named documents. Each document is interpreted as a literal and is associated with an IRI, thus forming what we call a *literalset*. Below is an example of a SPARQL-Generate query,[2] more complex examples can be found online.[3]

Document airport.csv	SPARQL-Generate request

Document airport.csv
```
id,stop,latitude,longitude
6523,25,50.901389,4.484444
7000,40,56.901389,4.584444
```

Output (in turtle)
```
<http://ex.com/6523>
  a transit:Stop;
  transit:route 21;
  geo:lat 50.901389;
  geo:long 4.484444 .
```

SPARQL-Generate request
```
GENERATE {
   ?airport a transit:Stop;
         transit:route ?route;
         geo:lat ?lat;
         geo:long ?long . }
SOURCE <http://ex.com/airport.csv> AS ?source
ITERATOR iter:CSV(?source) AS ?busStop
WHERE {
     BIND( fn:CSV(?busStop, "id" ) AS ?id )
     BIND( xsd:int( fn:CSV(?busStop, "stop") ) AS ?route)
     BIND( fn:CSV(?busStop, "longitude" ) AS ?long )
     BIND( fn:CSV(?busStop, "latitude" ) AS ?lat )
     BIND (URI(CONCAT("http://ex.com/",?id)) AS ?airport)
     FILTER( ?route < 30 ) }
```

Implementation on top of Apache Jena. SPARQL-Generate has been implemented on top of Apache Jena. It is available as open-source on GitHub.[4] It can be used as a Maven dependency, via a Web API, via a web form that itself uses the Web API, or as an executable jar. All of these tools may be found on the demonstration website:

http://w3id.org/sparql-generate

Supported data formats, and extensibility. Binding and iterator functions are available for the following formats: XML (exploiting XPath), CSV, TSV (conforming to the RFC 4180, or custom), HTML (exploiting CSS3 selectors), JSON and CBOR (exploiting JSONPath), and plain text (exploiting regular expressions). A complete documentation of the available binding and iterator functions is available at http://w3id.org/sparql-generate/functions.html.

The implementation relies on Jena's SPARQL binding function extension mechanism, and copies it for iterator functions. Therefore, covering a new data format in this implementation merely consists in implementing new binding and iterator functions in Jena. Even what is not covered by existing query languages can be implemented as an iterator function. For instance, iterator

---

[2] Prefixes are omitted to save space.

[3] http://w3id.org/sparql-generate/tests-reports.html.

[4] https://github.com/thesmartenergy/sparql-generate.

function `iter:JSONListKeys` iterates on key names of a JSON object, which is not feasible using JSONPath.

4   Future Work and Conclusion

Future plans consist of implementing more functions for more data formats, enabling on-the-fly function integration with an approach similar to [4], and adding some syntactic sugars that could strongly improve readability and conciseness of the queries. For instance one could use binding functions directly in the generate pattern, or use curly-bracket expressions instead of concatenating literals. Using such techniques, the example query from Sect. 3 could be shortened as follows:

```
GENERATE {
  <http://ex.com/{?id}> a transit:Stop;
      transit:route ?route ;
      geo:lat xsd:decimal( fn:CSV(?busStop, "latitude" ) );
      geo:long xsd:decimal( fn:CSV(?busStop, "longitude" ) ). }
SOURCE <http://ex.com/airport.csv> AS ?source
ITERATOR iter:CSV(?source) AS ?busStop
WHERE {
    BIND( fn:CSV(?busStop, "id" ) AS ?id )
    BIND( xsd:int( fn:CSV(?busStop, "stop") ) as ?route )
    FILTER( ?route < 30 ) }
```

The problem of exploiting data from heterogeneous sources and formats is common on the Web, and Semantic Web technologies can help in this regard. We introduced SPARQL-Generate, that extends the SPARQL language in its ability to generate RDF graphs, such that any non-RDF data sources, as well as RDF sources, can be exploited to create an output RDF graph. Its syntax closely matches SPARQL with little additions, and hence combines the following advantages: (i) it may be implemented on existing SPARQL engines; (ii) it is modular since extensions to new formats do not require a redefinition of the language (thanks to the use of SPARQL custom functions); and (iii) it is easy to learn by Semantic Web specialists that know SPARQL 1.1. Our implementation on top of Apache Jena covers many use cases, as reported on the dedicated web site http://w3id.org/sparql-generate.

References

1. Connolly, D.: Gleaning Resource Descriptions from Dialects of Languages (GRDDL). W3C Recommendation (2007). http://www.w3.org/TR/2007/REC-grddl-20070911/
2. Das, S., Sundara, S., Cyganiak, R.: R2RML: RDB to RDF Mapping Language. W3C Recommendation (2012). http://www.w3.org/TR/2012/REC-r2rml-20120927/
3. Dimou, A., Sande, M.V., Colpaert, P., Verborgh, R., Mannens, E., Van de Walle, R.: RML: a generic language for integrated RDF mappings of heterogeneous data. In: Proceedings of the Workshop on Linked Data on the Web (LDOW 2014) (2014)

4. Lefrançois, M., Zimmermann, A.: Supporting arbitrary custom datatypes in RDF and SPARQL. In: Sack, H., Blomqvist, E., d'Aquin, M., Ghidini, C., Ponzetto, S.P., Lange, C. (eds.) ESWC 2016. LNCS, vol. 9678, pp. 371–386. Springer, Cham (2016). doi:10.1007/978-3-319-34129-3_23

5. Polleres, A., Krennwallner, T., Lopes, N., Kopecký, J., Decker, S.: XSPARQL Language Specification. W3C Member Submission. http://www.w3.org/Submission/2009/SUBM-xsparql-language-specification-20090120/

6. Tandy, J., Herman, I., Kellogg, G.: Generating RDF from Tabular Data on the Web. W3C Recommendation. http://www.w3.org/TR/2015/REC-csv2rdf-20151217/

DOREMUS to Schema.org: Mapping a Complex Vocabulary to a Simpler One

Pasquale Lisena[(✉)] and Raphaël Troncy[(✉)]

EURECOM, Sophia Antipolis, France
{pasquale.lisena,raphael.troncy}@eurecom.fr

Abstract. Librarians and music professionals often us complex models and ontologies such as FRBRoo to represent music metadata. As a consequence, this metadata is not easily consumable by general search engines or external web applications. This paper presents a methodology, composed of a set of recipes, for mapping a complex ontology to a simpler model, namely Schema.org.

1 Music Information and Structured Data

Search engines and web applications display more and more knowledge panels alongside raw search results. This is made possible thanks to the adoption of some form of *Structured Data markup*, like Schema.org [4] that allows to describe the content of a web page in a machine-understandable way. In particular, Schema.org offers classes such as `CreativeWork` or `Event` and music-specific subclasses such as `MusicComposition` or `MusicEvent`.

Musical works are complex objects that require complex ontologies for expressing the richness of the information. FRBRoo is an ontology for describing bibliographic information [2]. The central feature of this ontology is the presence of the triplet Work-Event-Expression which considers that any artistic Work, only exists through an Event of creation, that realizes the Work itself into an Expression. The DOREMUS ontology[1] [1] extends FRBRoo with music-specific classes and properties like the key, the genre or the casting that could interpret a work. Because of this complexity, the consumption of these data by search engines and external applications is not easy. Nogales et al. proposed a mapping between the classes and properties of Schema.org and the ones of different vocabularies which have exactly the same name (or synonyms) [5]. Godby proposes to map each level of the chain Work - Expression - Manifestation - Item with a entity of the CreativeWork class [3]. A limit of this strategy is that the information is split up among different objects.

In this paper, we provide a methodology composed of four recipes for translating a complex ontology like DOREMUS into Schema.org. This method is based on the observation of the graph and it assumes sufficient knowledge of the source and target models. As example, we will represent Beethoven's *Sonata "Quasi una Fantasia"*, described in DOREMUS in Fig. 1 using Schema.org[2].

---

[1] http://www.doremus.org.
[2] In the following we use respectively the prefixes `mus:` and `sdo:`.

© Springer International Publishing AG 2017
P. Ciancarini et al. (Eds.): EKAW 2016 Satellite Events, LNAI 10180, pp. 136–139, 2017.
DOI: 10.1007/978-3-319-58694-6_17

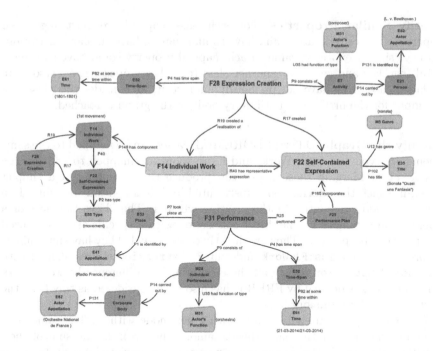

Fig. 1. Graph describing Beethoven's *Sonata "Quasi una Fantasia"* in DOREMUS.

2   Model Simplification

Choose the Starting Node. The most suitable starting point should coincide with the most significant class (or group of classes) in the source ontology (e.g. DOREMUS). It could be the class with the biggest number of instances, or it could consist in a frequent pattern, like the FRBRoo triangle. We choose *mus:F2 Expression* and *mus:F28 Expression Creation*, because they are linked with most of the crucial information for end-users such as the title and the composer.

Identify Similar Classes. For each class in DOREMUS, the best match in Schema.org should satisfy one or more of these criteria: 1. Have similar names, where similarity is computed using the Levenshtein distance or the number of common synonyms; 2. Have similar descriptions, where the similarity can be computed using the cosine or a Token-Wise distance; 3. Have similar properties; 4. Have similar expected property values (e.g. both `mus:U12_has_genre` and `sdo:musicCompositionForm` have "sonata" as possible value). If the search for a suitable match fails for a specific class, then it could mean that such an equivalent concept does not exist in Schema.org. In this case, this information can either be considered as too specific to be mapped or represented in a plain text note. Alternatively, new classes and properties could be considered in Schema.org extensions.

Identify Similar Properties. For each class mapped, we must align their properties. The criteria are again: have similar names, have similar descriptions and have similar expected values. Each mapped property could have as value a literal (e.g. key and genre) or another class (e.g. the composer is a Person). In the latter case, if we have not previously mapped it, we consider this class as a new input for the first steps, until every node in the graph is reached.

Simplify the Graph. Different DOREMUS classes can be mapped to the same Schema.org type: e.g. both Works and Expressions are mapped to sdo:Music-Composition. Merging these nodes can produce the advantage of a simpler model, in which the information is distributed in as less nodes as possible. Two nodes are good candidates for the merging when: 1. They have the class or a superclass in common; 2. The direct connection with a specific class is realized with the same property; 3. They are directly connected; 4. They have no conflicting properties. The mus:F1_Work and mus:F2_Expression classes aligned with sdo:MusicComposition satisfy all these criteria. The difference between Work and Expression as defined by FRBRoo has also been considered as tiny from the point of view of Schema.org [3].

Redundant nodes are substituted with a new node with: 1. The same class of the original one (or the most specific among the two); 2. The sum of their properties. The result of this phase is a new graph as depicted in Fig. 2.

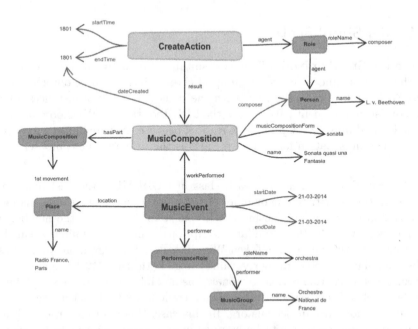

Fig. 2. Graph of Beethoven's *Sonata "Quasi una Fantasia"* in Schema.org.

3   Evaluation and Future Work

The evaluation of the goodness of the mapping is a long term goal, that can be realized only when data will be more easily consumed (e.g. by search engines). For having a quick feedback, we prepared a JSON-LD version of the *Sonata Quasi Una Fantasia* available at https://goo.gl/zzXVVh. Both the Structured Data Testing Tool by Google[3] and the Structured Data Linter[4] show the structure of the Schema.org graph like a tree, with 76 statements correctly recognized. The Structured Data Linter offers in addition a preview of the result as they could be shown in a SERP. For having a stronger visual feedback, we developed a lightweight web application, named *Schema.org Visualizer*[5], which consumes the data in JSON-LD and shows a knowledge cards for the described content. The result for our example is available at https://goo.gl/mCAszw.

We are planning to test our recipes on the ELI ontology[6] which is also based on FRBRoo. The goal is to compare the result of the mapping to the one which has recently been hand-made designed for being considered as a Schema.org extension[7]. Future work includes an implementation strategy for this methodology, that will also highlight the content excluded from the mapping, in order to better evaluate the possibility of presenting a Schema.org extension.

Acknowledgments. This work has been partially supported by the French National Research Agency (ANR) within the DOREMUS Project, under grant number ANR-14-CE24-0020.

References

1. Achichi, M., Bailly, R., Cecconi, C., Destandau, M., Todorov, K., Troncy, R.: DORE-MUS: Doing Reusable Musical Data. In: 14th International Semantic Web Conference (ISWC) (2015)
2. Doerr, M., Bekiari, C., LeBoeuf, P.: FRBRoo: a conceptual model for performing arts. In: Annual Conference of CIDOC, pp. 6–18 (2008)
3. Godby, C.J.: The Relationship between BIBFRAME and OCLC's Linked-Data Model of Bibliographic Description: A Working Paper. OCLC Research (2013)
4. Guha, R., Brickley, D., MacBeth, S.: Schema.org: evolution of structured data on the web. Commun. ACM **13**(9), 44–51 (2016)
5. Nogales, A., Sicilia, M.-A., Sánchez-Alonso, S., Garcia-Barriocanal, E.: Linking from Schema. org microdata to the Web of Linked Data: An empirical assessment. Comput. Stand. Interfaces **45**, 90–99 (2016)

---

[3] https://search.google.com/structured-data/testing-tool.
[4] http://linter.structured-data.org/.
[5] https://github.com/pasqLisena/schema-visualizer.
[6] http://publications.europa.eu/mdr/eli/.
[7] https://github.com/schemaorg/schemaorg/issues/1156.

FORMULIS: Dynamic Form-Based Interface for Guided Knowledge Graph Authoring

Pierre Maillot[1]([⊠]), Sébastien Ferré[1], Peggy Cellier[2], Mireille Ducassé[2], and Franck Partouche[3]

[1] IRISA/Université de Rennes 1,
Campus de Beaulieu, 35042 Rennes Cedex, France
{pierre.maillot,sebastien.ferre}@irisa.fr
[2] IRISA/INSA Rennes,
Campus de Beaulieu, 35042 Rennes Cedex, France
{peggy.cellier,mireille.ducasse}@irisa.fr
[3] IRCGN, 5 Boulevard Hautil, 95000 Cergy, France
franck.partouche@gendarmerie.interieur.gouv.fr

Abstract. Knowledge acquisition is a central issue of the Semantic Web. Knowledge cannot always be automatically extracted from existing data, thus contributors have to make efforts to create new data. In this paper, we propose FORMULIS, a dynamic form-based interface designed to make RDF data authoring easier. FORMULIS guides contributors through the creation of RDF data by suggesting fields and values according to the previously filled fields and the previously created resources.

1 Introduction

Manual acquisition of RDF data requires an effort either from contributors or from Semantic Web experts. In systems based on contributors' efforts, such as WebProtégé [4], the contributors have to learn Semantic Web principles, and how to use the specialized interfaces based on these principles. In systems based on Semantic Web experts' efforts, such as ActiveRaUL [1] or OntoWiki [2], the data input interfaces are easier to use, but have to be set up by the experts. In case of data evolution, contributors depend on experts to adapt the interface.

In this paper, we propose FORMULIS, a system that offers the user-friendliness of forms while guiding users with dynamic suggestions based on existing data. The contribution is threefold. First, the forms can be nested as deeply as necessary. Hence, it is possible to create at once several related entities, linked by property paths, by building nested descriptions. Second, the forms need no set up by experts because their content is based on SEWELIS suggestions [3]. Fields and values are suggested by taking into account all the fields and values already entered in the base and in the current form, hence offering dynamic and accurate

P. Maillot and S. Ferré—This work is partially supported by ANR project IDFRAud (ANR-14-CE28-0012).

P. Ciancarini et al. (Eds.): EKAW 2016 Satellite Events, LNAI 10180, pp. 140–144, 2017.
DOI: 10.1007/978-3-319-58694-6_18

suggestions. Last but not least, the forms can be extended with new fields and new sub-forms at any time, according to user needs and data evolution, with a simple click. New forms can also be created for new classes of resources.

FORMULIS has been used in a real setting. Six forensic experts from the forged ID unit of the document department of IRCGN[1] put into a knowledge base the description of thirty five forged Portuguese ID cards seized during a police operation. That experiment was successful despite experts having no knowledge of the Semantic Web, different level of expertise in forged ID detection, and different specialties. All were able to produce good quality data.

2   FORMULIS: Guided RDF Authoring Through Forms

FORMULIS aims at facilitating RDF authoring without knowledge of RDF by proposing a familiar interface, forms, and by dynamically suggesting values from current and previous inputs. On the one hand, graph-based data representations tend to become quickly illegible with the size of the graph and text-based notations, such as Turtle, need prior knowledge to be understood. On the other hand, forms are a common interface for data input. FORMULIS generates a data-driven form-based interface for the production of RDF data.

Interaction loop. At each iteration, the system computes suggestions to fill the form further, and the user selects one suggestion. The form is initially empty, and it is progressively filled through interaction steps, until it is judged complete by the user. Suggestions are computed with the help of the query relaxation and query evaluation mechanisms of SEWELIS [3]. This requires first to translate the partially-filled form into an initial query that is supposed to retrieve the possible values for the field under focus, given the already filled fields. Query relaxation of that initial query is in general necessary to ensure the existence of results. In order to propose legible suggestions, those results have then to be rendered as form elements, i.e. user interface widgets and controls. Finally, each time a user selects an element or activates a control, the form is modified by filling a field, adding a field, inserting a nested form, etc. From there, a new interaction step can start.

Nested forms and their RDF counterpart. A form F in FORMULIS is composed of an RDF class C, a resource URI R, and a set of RDF properties $p_1, \ldots, p_n$, along with their sets of values $V_1, \ldots, V_n$. Class C (e.g., `dbo:-Film`) determines the type of the resource being described, and resource R (e.g., `dbr:Corpse_Bride`) determines its identity. Each property p_i (e.g., `dbo:director`) defines a field of the form. Each field may have one or several values. Indeed, unlike tabular data, the RDF data model allows a resource to have several values for the same property, and it is therefore important to account for this in RDF editing tools. An example about films is the property `dbo:starring` relating films to actors. A field value can be one of: void, i.e. the field has not yet been filled; a literal, e.g. a string; an RDF resource already present in the knowledge

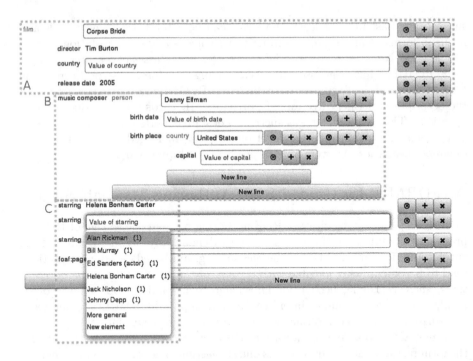

Fig. 1. Screenshot of FORMULIS during the creation of a film directed by Tim Burton using the DBpedia vocabulary, with parts in dotted frames for explanation purposes.

base (e.g., `dbr:Tim_Burton`); or a nested form to create and describe a new RDF resource. Those definitions of forms and field values exhibit the recursive nature of forms in FORMULIS, which allows for cascading descriptions of resources.

User interface. Fig. 1 shows an example of nested form, as it is displayed in the user interface of FORMULIS. In this example, the form is used to create the description of the film "Corpse Bride". The first step done by the user was to choose "Film" among the available classes in order to get the initial form to create a new film. The suggested fields of the initial form are based on the description of existing films in the base. The user interface follows the classical layout of a form with each field labeled by a property of the described resource. In frame (A), the "director" and "release date" are filled with RDF resources, respectively the URI labeled "Tim Burton" and the literal "2005". Note that RDFS labels are used to display URI in forms. Frame (B) shows two nested forms: one for the description of a new person as "music composer" of the current film ("Danny Elfman"), and another for the description of his birth place ("United States"). In frame (C), the resource labeled "Helena Bonham Carter" has been entered as starring in the movie. In another field for the "starring" property, the user is given six suggestions corresponding to actors that have been starring in movies similar to "Corpse Bride", either because they were also directed by "Tim Burton", released in 2005 and/or starring "Helena Bonham

Carter". Below those suggestions, the user is also given the possibility to create a new resource ("New element"), or to ask for more suggestions by not taking into account the previously filled fields ("More general").

Suggestions. In order to suggest values for a field, we need to translate the form to a SPARQL query that retrieves existing values for that field in the knowledge base. Suppose that the user has set the focus on the "starring" field in Fig. 1 because she wants to get suggestions for the actors of the film. We could retrieve all values of the "starring" property in the base, using query: `SELECT ?v WHERE {?x dbo:starring ?v}`. However, this would make suggestions unspecific. We make suggestions dynamic and more accurate by taking into account all filled fields. The principle is to generate a SPARQL query whose graph pattern is the current description: `SELECT ?v WHERE {?x a dbo:Film; dbo:director dbr:Tim_Burton; dbo:releaseDate "2005"; dbo:musicComposer [a dbo:Person; dbo:birthPlace [a dbo:Country]]; dbo:starring dbr:Helena_B_Carter, ?v.}`. That query is sent to SEWELIS for relaxation and evaluation.

Cold start. In the case of an empty base or a new class, the creation of the first instance of the new class requires the creation of a new form and its fields, in addition to entering field values. That first instance is enough to generate the forms for the following instances. The user can also extend generated forms with new fields, as it can be seen in Fig. 1 at the bottom of each form ("New line").

3   Demo

The demo will show that the accuracy of suggestions improves when more fields are filled, and when more resources are put in the knowledge base. That accuracy will be shown by the suggestions of values reflecting patterns commonly found in data, for example the fact that actor Johnny Depp is often starring with Helena B. Carter in films directed by Tim Burton. Participants will be given the opportunity to enter the description of recent movies, starting from DBpedia films. The demo will also show the cold start case with the creation of new forms. FORMULIS is available in beta version at http://servolis.irisa.fr:8080/formulis/.

References

1. Butt, A.S., Haller, A., Liu, S., Xie, L.: ActiveRaUL: A Web form-based User Interface to create and maintain RDF data. In: International Semantic Web Conference, Posters & Demonstrations Track, vol. 1035, pp. 117–120 (2013)
2. Frischmuth, P., Martin, M., Tramp, S., Riechert, T., Auer, S.: OntoWiki - an authoring, publication and visualization interface for the data web. Semant. Web **6**(3), 215–240 (2015)

3. Hermann, A., Ferré, S., Ducassé, M.: An interactive guidance process supporting consistent updates of RDFS graphs. In: Teije, A., Völker, J., Handschuh, S., Stuckenschmidt, H., d'Acquin, M., Nikolov, A., Aussenac-Gilles, N., Hernandez, N. (eds.) EKAW 2012. LNCS, vol. 7603, pp. 185–199. Springer, Heidelberg (2012). doi:10.1007/978-3-642-33876-2_18

4. Tudorache, T., Nyulas, C.: Natalya F Noy, and Mark A Musen: WebProtégé: A collaborative ontology editor and knowledge acquisition tool for the web. Semantic web 4(1), 89–99 (2013)

Usability and Improvement of Existing Alignments: The LOINC-SNOMED CT Case Study

Mélissa Mary[1,2(✉)], Lina Soualmia[2,3], and Xavier Gansel[1]

[1] bioMérieux S.A., La Balme Les Grottes, France
{melissa.mary,xavier.gansel}@biomerieux.com
[2] Normandy University, LITIS EA 4108 and NormaSTIC CNRS 3638, Rouen, France
Lina.Soualmia@chu-rouen.fr
[3] French National Institutes for Health (INSERM), LIMICS UMR_1142, Paris, France
http://www.biomerieux.com

Abstract. LOINC® and SNOMED CT® are two of the most used biomedical terminology standards to conjointly describe medical laboratory data into patient Electronic Health Records. The institutions owning them entered in a collaboration 4 years ago. The intention was to provide alignments between LOINC® and SNOMED CT® in order to improve query and aggregation of patient data. This work brings input on the LOINC—SNOMED CT alignment effort: (i) we developed algorithms aiding to align LOINC® and SNOMED CT® efficiently and (ii) we demonstrated the benefits of the SNOMED CT® conceptual hierarchy and tests model to query data initially coded in LOINC®.

Keywords: Biomedical terminology · LOINC · SNOMED CT · Ontology alignment · Ontology reuse · Reasoning

1 Introduction

In several countries a trend in health IT brings the centralization of patient medical data into Electronic Health Records (EHR). Semantic interoperability of those data and systems requires the use of standard vocabularies called biomedical terminologies. In the case of laboratory medical data reporting, regulatory bodies and health IT standards recommend the LOINC terminology [1] to describe tests performed and the SNOMED CT ontology [2] to represent test results. To support data analytics and information retrieval in EHR, laboratory test results have to be contextualized with information describing the tests. The two institutes owning SNOMED CT® (IHTSDO) and LOINC® (Regenstrief Institute) started a collaboration several years ago [3, 4] to interoperate the two resources so that LOINC® tests can be described as SNOMED CT® post-coordinated expressions.

In this study, we first proposed automatic alignment solutions that (i) increment existing alignments and (ii) ensure consistency and subsequently may reduce the human expertise required. Secondly we analyze the proposed mapping to quantify the benefits and issues of using SNOMED CT® expressions to represent LOINC® codes.

© Springer International Publishing AG 2017
P. Ciancarini et al. (Eds.): EKAW 2016 Satellite Events, LNAI 10180, pp. 145–148, 2017.
DOI: 10.1007/978-3-319-58694-6_19

2 Material and Method

2.1 LOINC—SNOMED CT Alignment

The third version of LOINC—SNOMED CT alignment represents 20% of LOINC codes (13,786 tests) into SNOMED CT post-coordinate expressions [3, 4]. Test description in LOINC is structured by six feature as System to represent sample used for tests. LOINC—SNOMED CT also represent:

- the alignment of 4050 parts (feature value as urine) and concept;
- a new model to describe tests (System feature is represented by two attributes).

2.2 LOINC Test Classification with SNOMED CT Ontology

The test classification process is a two-step procedure. First of all, LOINC—SNOMED CT OWL file is merged with SNOMED CT January 2016 international release. Secondly, ELK reasoner [5], which allows reasoning on OWL EL large ontologies, is used to reclassify the merged ontologies.

2.3 Alignment Process

Our alignment strategy ('Anchor Flooding') is derived from the algorithm developed by Seddiqui and Aono [6]. This is a recursive and heuristic alignment approach that extends a pre-existing alignment (initial anchors) between LOINC and SNOMED CT by comparing taxonomical related concepts. Alignment used both lexical and semantic approaches. Lexical approach (Eq. 1) consists in Stoilos similarity [7] and our implementation of bag of words similarity (called WGram). We implemented a semantic approach (A_{sem}) which is based on UMLS Metathesaurus [8] to retrieve the term meanings. Semantic alignments are characterized by two metrics (i) the similarity (sim_{sem}) (ii) the confidence measured between terms and their semantic representation ($conf_{sem}$).

New anchors were selected (Eq. 2) using thresholds parameters:

- $t_{St-WGram} = 0.75$ as threshold for lexical alignments,
- $t_{sem} = 0.75$ to filter semantic alignment according their similarity and
- $t_{conf} = 600$ threshold which filter semantic alignment according their confidence measure. Each threshold we used were optimized beforehand.

$$A_{St-WGram}\left(T_1, T_2 | t\right) := A_{St}\left(T_1, T_2 | t\right) \cap A_{WGram}\left(T_1, T_2 | t\right) \tag{1}$$

$$A\left(T_1, T_2 \middle| t_{St-WGram}, t_{sem}, t_{conf}\right) := A_{sem}\left(T_1, T_2 \middle| t_{sem}, t_{conf}\right) \cap \\ \left(A_{St-WGram}\left(T_1, T_2 | t_{St-WGram}\right) \cap A_{sem}\left(T_1, T_2 \middle| t_{conf} = 0\right)\right) \tag{2}$$

3   Results

3.1   Enhancing Alignment Using Automatic Methods

We randomly selected 68 alignments in the mapping part Table (3%) to be used as initial anchors. We observed that about 30% of initial anchor (20) retrieves 192 new anchors. During the alignment review (Table 1), distinction was made on the alignments between the hierarchically related parts and concepts (Broader to Narrower terms, *i.e. keratin* aligned with *beta keratin*). They are caused by differences between classification of parts (LOINC) and concepts (SNOMED CT). The curation highlighted that semantic and lexical metrics are complementary. We proposed a post-alignment filter in order to increase precision (Eq. 3). BestForBoth is a contextual filter that selects alignments only if alignment between a part and a concept is the best in both directions.

$$\text{Filter}(\mathcal{A}) := \text{BestForBoth}(\mathcal{A}_{st-WGram}) \cup \text{Filter}(\mathcal{A}_{sem} \big| t_{conf} > 800) \qquad (3)$$

Table 1. Curation of new anchors according metrics used in alignment process.

Curation	Total (192)	Lexical al. (66)	Semantic al. (93)	Both al. (33)	Filter($\mathcal{A}$) Eq. 3 (137)
FALSE	136 (70%)	49 (76%)	68 (73%)	19 (57.5%)	85 (59.5%)
Broader to Narrower term	25 (13%)	9 (13%)	14 (15%)	2 (6%)	21 (15.3%)
TRUE	29 (16%)	7 (10%)	10 (11%)	12 (36.5%)	29 (21%)
Unknown	2 (1%)	1 (1%)	1 (1%)	0	2 (1%)

3.2   Usability of SNOMED CT Tests Classification

We observed SNOMED CT enables to classify 50% of tests (Table 2). The use of Description Logic improves the expressivity of queries we performed. However we encountered issues with SNOMED CT to build simple query that aims to retrieve all

Table 2. Comparison of LOINC and SNOMED CT to classify tests and express query.

Data (Reasoner)	LOINC (None)	Alignment + SNOMED CT® (ELK)
Number of tests	13,756	
Number of tests asserted as equivalent	0	45
Number of tests subsumed by another test	0	6,789
Query language	Keyword	Description Logic
Type of result	Match strictly	Match and hierarchically related
Usability for simple query	Easy	Require SNOMED CT expertise in (i) test model (ii) semantic
Usability for complex query	Difficult	Easy

tests performed on *blood* sample. In SNOMED CT there are three concepts to represent blood. To understand the distinction of the three concepts, the user should have background knowledge in SNOMED CT. Even if LOINC terminology is not adapted to express complex queries (composed by multiple restrictions), it is more intuitive and user friendly than SNOMED CT.

4   Conclusion

The results we obtained in expanding LOINC SNOMED CT alignments highlight the importance of taking into account both lexical and semantic dimensions. However, the parameter we used doesn't give results with good sensitivity and it still has to be refined with a post alignment filter. We also showed that using SNOMED CT concept classification can improve the test classification initially proposed by LOINC. The comparison of the query expression demonstrated the benefits of using SNOMED CT even if the creation of a query into SNOMED CT requires skills on the SNOMED CT structure.

References

1. Huff, S.M., Rocha, R.A., McDonald, C.J., De Moor, G.J.E., Fiers, T., Bidgood, W.D., Forrey, A.W., Francis, W.G., Tracy, W.R., Leavelle, D., Stalling, F., Griffin, B., Maloney, P., Leland, D., Charles, L., Hutchins, K., Baenziger, J.: Development of the logical observation identifier names and codes (LOINC) vocabulary. J. Am. Med. Inform. Assoc. JAMIA **5**, 276–292 (1998)
2. Bhattacharyya, S.B.: Using SNOMED CT. In: Bhattacharyya, S.B. (ed.) Introduction to SNOMED CT, pp. 157–182. Springer, Singapore (2016)
3. IHTSDO. Regenstrief Institute: Regenstrief and the IHTSDO are working together to link LOINC and SNOMED CT (2013). https://loinc.org/collaboration/ihtsdo
4. Vreeman, D.: Guidelines for using LOINC and SNOMED CT together (2015). https://danielvreeman.com/guidelines-for-using-loinc-and-snomed-ct-together-without-overlap/
5. Kazakov, Y., Krötzsch, M., Simančík, F.: ELK: a reasoner for OWL EL ontologies. System Description (2012)
6. Seddiqui, M.H., Aono, M.: An efficient and scalable algorithm for segmented alignment of ontologies of arbitrary size. Web Semant. Sci. Serv. Agents World Wide Web **7**, 344–356 (2009)
7. Stoilos, G., Stamou, G., Kollias, S.: A string metric for ontology alignment. In: Gil, Y., Motta, E., Benjamins, V.R., Musen, M.A. (eds.) ISWC 2005. LNCS, vol. 3729, pp. 624–637. Springer, Heidelberg (2005). doi:10.1007/11574620_45
8. Schuyler, P.L., Hole, W.T., Tuttle, M.S., Sherertz, D.D.: The UMLS metathesaurus: representing different views of biomedical concepts. Bull. Med. Libr. Assoc. **81**, 217–222 (1993)

LD Sniffer: A Quality Assessment Tool
for Measuring the Accessibility of Linked Data

Nandana Mihindukulasooriya[✉], Raúl García-Castro,
and Asunción Gómez-Pérez

Ontology Engineering Group,
Universidad Politécnica de Madrid, Madrid, Spain
{nmihindu,rgarcia,asun}@fi.upm.es

Abstract. During the last decade, the Linked Open Data cloud has grown with much enthusiasm and a lot organizations are publishing their data as Linked Data. However, it is not evident whether enough efforts have been invested in maintaining those data or ensuring their quality. Data quality, defined as *"fitness for use"*, is an important aspect for Linked Data to be useful. Data consumers use quality indicators to decide whether or not to use a dataset in a given use case, which makes quality assessment of Linked Data an important activity. Accessibility, which is defined as *the degree to which the data can be accessed*, is a highly relevant quality characteristic to achieve the benefits of Linked Data. In this demo paper presents LD Sniffer, a web-based open source tool for performing quality assessment on the accessibility of Linked Data. It generates unambiguous and comparable assessment results with explicit semantics by defining both quality metrics as well as assessment results in RDF using the W3C Data Quality vocabulary. LD-Sniffer is also distributed as a Docker image improving ease of use with zero configurations.

Keywords: Linked data · Quality · Accessibility · DQV

1 Introduction

The Linked Data[1] principles promote publishing data in a machine-readable manner using Web standards. The third Linked Data principle, which states that *"When someone looks up a URI, provide useful information, using the standards"*, and the fourth one, which says *"Include links to other URIs so that they can discover more things"*, are related to the accessibility of Linked Data resources. Links to related entities in Linked Data make it possible to start from one piece of data and traverse through different sources with a follow-your-nose approach[2] in order to discover more entities and get contextual information. This virtue of Linked Data makes it possible to treat the Web as a global data

---

[1] https://www.w3.org/DesignIssues/LinkedData.html.
[2] http://patterns.dataincubator.org/book/follow-your-nose.html.

© Springer International Publishing AG 2017
P. Ciancarini et al. (Eds.): EKAW 2016 Satellite Events, LNAI 10180, pp. 149–152, 2017.
DOI: 10.1007/978-3-319-58694-6_20

space with typed links between data from different sources breaking isolated data silos [1]. However, accessibility plays a key role to materialize this vision.

Accessibility is defined as the degree to which the data can be accessed or the proper functioning of access methods so that it is obtainable by a consumer [2]. In Linked Data, accessibility can be checked by dereferencing resource IRIs (which should be HTTP URIs according to the Linked Data principles). The existing links may become not dereferenceable due to many reasons such as not being maintained, server failures, domain expiration, or typos in the URIs. The tool presented in this paper, LD Sniffer, uses several metrics based on the dereferenceability of URIs in Linked Data to derive quality indicators.

2   Related Work

Zaveri et al. [3] provided a survey of different quality dimensions and metrics found in the literature. Besides, there are several tools in the literature that support quality assessment of Linked Data according to such metrics. On the one hand, there are tools that are focused on assessing quality along one quality dimension, such as trust (TrustBot, Trellis, tSPARQL) or interlinking (LinkedQA, LiQuate). On the other hand, there are tools that are frameworks for generic quality assessment such as Luzzu[3], RDFUnit[4], or Sieve[5]. Furthermore, there is a set of tools that allows an exploratory inspection of quality issues such as ProLOD[6], LODStats[7], ABSTAT[8], or Loupe[9] [4] that mainly use different statistics and patterns extracted from data. The LD Sniffer tool falls into the category of tools that assess quality along one dimension, in this case, accessibility.

3   LD Sniffer

LD Sniffer[10] is a tool that allows users to assess the quality of Linked Data resources in a dataset, such as DBpedia or Wikidata. Given a set of resource URIs, each identified by an IRI (for example, "http://es.dbpedia.org/resource/Spain"), it dereferences each IRI using the HTTP GET method and obtains the corresponding RDF graph. In that sense, it acts as a Linked Data crawler. Then, it analyzes all the IRIs in the given graph and assesses them based on the accessibility metrics defined in the Linked Data Quality Model (LDQM) [5].

LD Sniffer has been developed as an open source project under the Apache 2.0 license. It is written in Java 8 and it uses multithreading for efficiently analyzing

---

[3] http://eis-bonn.github.io/Luzzu/.
[4] http://aksw.org/Projects/RDFUnit.html.
[5] http://sieve.wbsg.de/.
[6] http://hpi.de/en/naumann/projects/data-profiling-and-analytics/prolod.html.
[7] http://stats.lod2.eu/.
[8] http://abstat.disco.unimib.it:8880/.
[9] http://loupe.linkeddata.es/.
[10] https://github.com/nandana/ld-sniffer.

a large number of IRIs. Furthermore, it uses caching to avoid evaluating the same IRI multiple times within a short period, and request delays and request throttling techniques to avoid overloading servers with too many requests. It provide a web interface and also available as a Docker image.

LD Sniffer allows both evaluating a single Linked Data resource or a collection of resources such as a dataset. It also allows the user to select which metrics to be evaluated in the quality assessment.

3.1 Metrics

An important aspect of designing a quality assessment tool is to decide the metrics for measuring a given quality characteristic. In this sense, LD Sniffer uses the LDQM quality model to define its metrics. The LDQM quality model provides a set of metrics that can be used to measure different quality characteristics of Linked Data and those categorized as base measures, derived measures, and indicators as shown in the LDQM quality model wiki[11].

The metrics in the quality model are defined as Linked Data with global identifiers and explicit semantics using the W3C Data Quality Vocabulary[12]. This allows the LD Sniffer tool to advertise the metrics that it uses in an unambiguous manner. Listing 1.1 illustrates an excerpt of a metric definition representation[13].

Listing 1.1. A metric definition using DQV and LDQM

```
indicator:AverageIrIderef a  dqv:Metric,  qmo:QualityIndicator;
    dc:description  "Average of  dereferenceable IRIs in a  triple,
        a graph,  or a dataset."@en;
    skos:prefLabel  "Average IRI  dereferenceability"@en;
    qmo:hasScale  scale:percentageHigherBest;
    qmo:measuresCharacteristic  dimension:Accessibility;
    ldq:hasAspect  ldq:LinkedDataServer;
    ldq:hasGranularity  ldq:Triple,  ldq:Graph ,  ldq:Dataset ;
    skos:definition  "(Number of  dereferenceable  IRIs/Number of
        distinct IRIs) * 100
    dqv:expectedDataType  xsd:double;
    dqv:inDimension  dimension:Accessibility;
    ldq:calculatedWith  technique:UrlDereferencing;
    prov:wasDerivedFrom  basemeasure:Numberofdistinctiris;
    derivedMeasure:Numberofdereferenceableiris  .
```

3.2 Quality Assessment Results

Another important aspect of the tool is the representation of the assessment results. LD Sniffer uses the W3C Data Quality Vocabulary to describe the evaluation results. In addition to the measurement values and their respective metrics, LD Sniffer provides a lot of provenance information such as when did the

---

[11] http://delicias.dia.fi.upm.es/LDQM/index.php/Accessibility.

[12] https://www.w3.org/TR/vocab-dqv/.

[13] Prefixes are omitted for brevity and are aligned with prefixes in http://prefix.cc/.

evaluation start and end, what techniques and tools were used, and whether the measurement values were estimated, etc. using vocabularies such as PROV-O[14] or EVAL[15]. The quality issues are reported using the Quality Problem Report[16] ontology. Sample evaluation results can be found in GitHub[17].

Furthermore, LD sniffer was used to perform an assessment of the accessibility of a subset of approximately 100 thousand DBpedia resources containing more than 1 million IRIs. The evaluation results are available both as a downloadable RDF dump[18] in datahub as well as in a SPARQL endpoint[19] for querying. They are also analyzed in the LD Sniffer website[20].

4   Conclusions and Future Work

This demo presents a command line tool that can be used to easily assess the accessibility of Linked Data resources. By (a) defining the metrics used in quality assessment in detail using a Linked Data Quality Model, (b) representing both metrics and assessment results in RDF using the W3C Data Quality Vocabulary, and (c) providing provenance information, LD Sniffer makes sure that quality results are comparable and can be used by other third parties.

As future work, we plan to extend the tool to assess other quality dimensions as well as to provide the tool as an online web application with a GUI.

Acknowledgments. This work was funded by the BES-2014-068449 grant under the 4V project (TIN2013-46238-C4-2-R).

References

1. Heath, T., Bizer, C.: Linked data: evolving the web into a global data space. Synth. Lectures Semant. Web Theory and Technol. **1**(1), 1–136 (2011)
2. Joint Technical Committee ISO/IEC JTC 1, Information technology, Software and System Engineering: ISO/IEC 25012 - Data Quality Model. Standard, ISO, Geneva, CH, December 2008
3. Zaveri, A., Rula, A., Maurino, A., Pietrobon, R., Lehmann, J., Auer, S.: Quality assessment for linked data: a survey. Semant. Web **7**(1), 63–93 (2015)
4. Mihindukulasooriya, N., Poveda-Villalón, M., García-Castro, R., Gómez-Pérez, A.: Loupe - an online tool for inspecting datasets in the linked data cloud. In: Demo at the 14th International Semantic Web Conference, Bethlehem, USA (2015)
5. Radulovic, F., Mihindukulasooriya, N., García-Castro, R., Pérez, A.G.: A comprehensive quality model for linked data. Semant. Web J. (2017)

---

[14] http://www.w3.org/TR/prov-o/.

[15] http://purl.org/net/EvaluationResult.

[16] http://purl.org/eis/vocab/qpro.

[17] http://nandana.github.io/ld-sniffer/examples/results.ttl.

[18] https://datahub.io/dataset/ldqm-dbpedia-2016.

[19] http://nandana.github.io/ld-sniffer/sparql.html.

[20] http://nandana.github.io/ld-sniffer/.

A Benchmarking Framework
for Stream Processors

Andreas Moßburger, Harald Beck, Minh Dao-Tran[✉], and Thomas Eiter

Institute of Information Systems, Vienna University of Technology,
Favoritenstraße 9-11, 1040 Vienna, Austria
{mossburger,beck,dao,eiter}@kr.tuwien.ac.at

1 Introduction

Stream Processing/Reasoning, an active research topic [5], has been picked up by different communities which developed a diversity of stream processors/reasoners. This however makes empirical evaluation and comparison of these engines a non-trivial task [4]. Different classes of those engines work on different formats of input data, use different languages to formulate queries, evaluate these queries using different semantics and produce different formats of output. To be able to compare such engines, a benchmarking framework that can cope with this wide diversity is needed.

This paper proposes a generic architecture for generating/gathering, streaming data, for evaluating different stream processors/reasoners and for running those evaluations.

We show how our framework can be used with the GTFS domain to evaluate the two representative RDF Stream Processing (RSP) engines C-SPARQL [3] and CQELS [7], and the Spark engine,[1] a powerful tool set for implementing stream processing applications that is widely used in Big Data. In that, we use the Answer Set Programming (ASP) engine clingo[2] as a base-line for comparing the results.

2 Challenges

The following challenges arised in comparing the aforementioned engines:

– The engines work on different data formats: RSP engines need data in some RDF serialization, Spark reads CSV files and clingo needs facts in ASP syntax.
– To provide not only fair comparisons but also reproducible tests, live streaming data has to be captured, so that it can be replayed in a reproducible manner.
– There is no unified way to query the engines: C-SPARQL and CQELS are queried by SPARQL-like queries, while Spark can be accessed by writing Scala functions using its API; clingo instead requires one to write logic programs.
– Output has to be generated in a unified format for easy comparison.

This research has been supported by the Austrian Science Fund (FWF) projects P24090, P26471, and W1255-N23.

[1] http://spark.apache.org/.
[2] https://github.com/potassco/clingo.

P. Ciancarini et al. (Eds.): EKAW 2016 Satellite Events, LNAI 10180, pp. 153–157, 2017.
DOI: 10.1007/978-3-319-58694-6_21

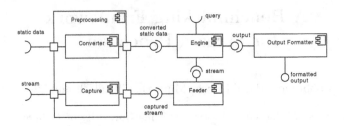

Fig. 1. Component diagram of modules

3   Framework Architecture

We propose a modular architecture composed of small components with clearly defined tasks and interfaces. Each module tackles one of the challenges described above. This architecture is generic enough to be implemented using a wide variety of technologies, from small scripts interacting together to classes in rather monolithic projects. To enable unified access to different engines, lightweight wrappers are utilized.

Figure 1 shows the modules of our architecture and the resulting data flow. The *converter* and *capture* module can be seen as two parts of a *preprocessing* module which is responsible for converting data to a usable format.

Converter is responsible for converting any static data from the provided format to a format that can be read by an engine. In some cases this module may even be omitted.

Capture extracts relevant data from a data stream and stores it, allowing reproducible evaluations. The format of the stored data should be generic, so that only minimal conversions are necessary for a particular engine. Additionally, timing information of the captured data should be stored, so it can be played back authentically.

Feeder is responsible for replaying the captured streaming data to the engines. It allows arbitrary fine control over the streaming process. Data may be streamed using authentic or artificial timing, like streaming a certain amount of data per time unit.

Engine wraps the evaluated engine. Wrappers provide a standardized way of accessing/outputting data for different engines but are not allowed to affect their performance.

Output Formatter converts the output data from different engines to a canonical form.

4   Show Case

4.1   General Transit Feed Specification

A public transport scenario offers ample opportunities for interesting data processing and reasoning tasks. In addition, the open data movement lead to

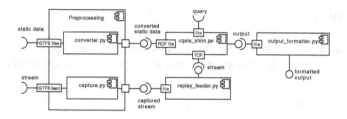

Fig. 2. Component diagram for GTFS and CQELS

free availability of public transport data for many cities worldwide. Most public transport data is published in the GTFS (General Transit Feed Specification) [1] format developed by Google and the Portland TriMet transit agency. A GTFS feed provides information suitable for trip planning. Moreover, the standard leaves enough room for extensions beyond these capabilities. While GTFS specifies static data useful for trip planning, its accompanying GTFS-realtime extension was developed to describe real-time data.

GTFS-realtime provides *TripUpdates* and *VehiclePositions*. A *TripUpdate* represents a change to a timetable and consists of possibly multiple delays or new arrival times for single stops of a trip. A *VehiclePosition* tells the position of a vehicle relative to a stop.

The public availability of real world data makes GTFS ideal for evaluating stream processing/reasoning engines.

4.2 Implementation

We implemented the components as Python scripts which communicate using files and TCP connections. This highly transparent working model allows to exchange single components easily and to examine intermediate results. By utilizing powerful and well documented libraries for handling GTFS and RDF data a low entry barrier is maintained. The following scripts are provided (our code is available at [2]):

- `gtfs-converter.py` and `gtfs-capture.py` implement the converter and capture modules, resp. These scripts are specific to the GTFS use case. All other scripts and programs are generic and do not make any assumptions about the data domain.
- `simple_feeder.py`, `replay_feeder.py` and `triple_to_asp.py` provide implementations of the feeder module.
- `output_formatter.py` covers the output formatter module.

For the engine module, different wrappers were implemented to access the engines. Figure 2 illustrates instantiating the architecture with scripts/programs to evaluate CQELS.

5   Evaluation

This section briefly summarizes the evaluation of the engine described above using our framework. All the queries used in this evaluation are available at [2].

5.1   Functionality

Our first evaluation was a basic test to determine which features of SPARQL 1.1 are supported by C-SPARQL and CQELS, namely FILTER, UNION, OPTIONAL, arithmetic, aggregation such as COUNT, COUNT DISTINCT, MAX, post-processing with ORDER BY, joining. We also tested which features could be replicated with Spark and clingo.

As a result, all mentioned features are supported by C-SPARQL and can be covered by clingo, CQELS lacks support for UNION, OPTIONAL and some aggregations. Spark misses out only ORDER BY.

5.2   Correctness

When comparing different engines, the underlying semantics cannot be ignored. Thus, one needs a highly expressive language to serve as an oracle for correct results via simulating input streams. We chose Answer Set Programming as it offers high expressiveness. Moreover, a number of mature ASP engines have been developed; among them clingo is of particular interest, as it showed top performance in many ASP competitions.

Note that C-SPARQL, CQELS and Spark generally output a multiset of tuples as result. Only in queries with an ORDER BY clause, the order of the output is not arbitrary. On the other hand, clingo always outputs a set as result. Because of this discrepancy one might argue that ASP is not suitable to serve as an oracle for comparing the other engines. However, if in practical queries multiple occurrence of a result matters, GROUP BY and COUNT clauses are used. If a result may occur multiple times but it is not of interest how often, DISTINCT is used. Therefore, for this fragment, the result will be a set and using ASP as an oracle is perfectly adequate.

In our run of the evaluation, the results of CQELS and Spark conformed to the results predicted by clingo. C-SPARQL produced correct results too, but was missing a few lines of output, compared to the others.

6   Conclusions

We presented a generic benchmarking framework that allows one to easily plug in her own stream processing/reasoning engine and compare it with stream engines developed by different communities. The evaluation is done on real public transportation data in GTFS format, a data domain that allows practical queries to test different interesting aspects of stream processing/reasoning. We provided a set of queries to test query functionalities and compared the results of C-SPARQL,

CQELS, and Spark using an oracle built on clingo. For future work, we plan to plug more engines into the framework and extend the query set to cover more stream processing/reasoning aspects.

References

1. General transit feed specification. https://developers.google.com/transit/. Accessed 15 Sep 2016
2. sr_data_generator   github   repository.   https://github.com/mosimos/sr_data_generator/. Accessed 15 Sep 2016
3. Barbieri, D.F., Braga, D., Ceri, S., Valle, E.D., Grossniklaus, M.: C-SPARQL: a continuous query language for RDF data streams. Int. J. Semant. Comput. **4**(1), 3–25 (2010)
4. Dao-Tran, M., Beck, H., Eiter, T.: Contrasting RDF stream processing semantics. In: Qi, G., Kozaki, K., Pan, J.Z., Yu, S. (eds.) JIST 2015. LNCS, vol. 9544, pp. 289–298. Springer, Cham (2016). doi:10.1007/978-3-319-31676-5_21
5. Valle, D.E., Ceri, S., van Harmelen, F., Fensel, D.: It's a streaming world! reasoning upon rapidly changing information. IEEE Intell. Syst. **24**, 83–89 (2009)
6. Gebser, M., Kaminski, R., Kaufmann, B., Schaub, T.: Clingo = ASP + Control: preliminary report. CoRR, abs/1405.3694 (2014)
7. Le-Phuoc, D., Dao-Tran, M., Xavier Parreira, J., Hauswirth, M.: A native and adaptive approach for unified processing of linked streams and linked data. In: Aroyo, L., Welty, C., Alani, H., Taylor, J., Bernstein, A., Kagal, L., Noy, N., Blomqvist, E. (eds.) ISWC 2011. LNCS, vol. 7031, pp. 370–388. Springer, Heidelberg (2011). doi:10.1007/978-3-642-25073-6_24

Life Science Ontologies in Literature Retrieval: A Comparison of Linked Data Sets for Use in Semantic Search on a Heterogeneous Corpus

Bernd Müller[✉], Alexandra Hagelstein, and Thomas Gübitz

German National Library of Medicine,
ZB MED - Information Centre for Life Sciences,
Gleueler Straße, 50931 Cologne, Germany
{bernd.mueller,hagelstein,guebitz}@zbmed.de
http://www.zbmed.de/

Abstract. Ontologies are modeled using specific concepts of the knowledge domain as well as using generic concepts. Life science ontologies like MeSH, `Agrovoc`, and `DrugBank` are helpful for searching through large corpora. The distinct linkage to either the agricultural domain or the medical domain cannot be resolved for generic concepts that were created when modeling the domain. In information retrieval, it is required to filter knowledge resources for domain specific concepts in order to avoid noise in search results caused by generic concepts. Here, we present an exploratory step towards evaluating concept frequencies amongst different knowledge domains when employing ontologies in the retrieval on a large corpus.

Keywords: Life sciences · Linked open data · Named entity recognition · Semantic search

1 Introduction

Finding relevant information in large electronic text collections such as digital libraries becomes increasingly difficult because of the increasing volume [1] and heterogeneous properties. The incorporation of ontologies for guiding the search in literature databases can be useful in order to improve the accuraccy of relevance in the retrieved information. The employment of a matching ontology is straightforward if the database is specific to the particular field of science. Contrastingly, if the database contains heterogeneous resources covering a range of scientific fields, more general ontologies or a combination of specific ontologies may be useful. The three major steps for chosing and employing ontologies in a retrieval environment are:

1. the selection process of one or more ontologies to expand the search terms to closely related concepts.
2. the generation of hit tables of search results according to semantic terms and their weights.

© Springer International Publishing AG 2017
P. Ciancarini et al. (Eds.): EKAW 2016 Satellite Events, LNAI 10180, pp. 158–161, 2017.
DOI: 10.1007/978-3-319-58694-6_22

3. the structuring of search results as a graph derived from co-occurring concepts in the ontology.

It is essential which ontologies are selected for enriching the query and for structuring the search results because the specificity of ontologies varies vastly. Here, three different and frequently used ontologies are compared in relation to their concepts occurring in LIVIVO[1].

2   Methodology

The literature search engine LIVIVO holds a corpus of 63 million citations having a focus on life sciences including the scientific fields of medicine, health, agriculture, nutrition and environment. In previous work, a UIMA[2]-based text and data mining workflow was employed to recognize life science concepts in the corpus of LIVIVO [5]. In this workflow, the analysis engine ConceptMapper [7] was used with dictionaries that were created from the Medical Subject Headings (MeSH)[3] [6], the pharmaceutical database DrugBank[4] [4], and the multilingual agricultural vocabulary AGROVOC[5] [3]. The results of the employed named entity recognition workflow are available online as linked dataset.[6] As overview, the frequency information of the found concepts is shown in Table 1. In order to measure the similarity between the three datasets, the Jaccard index is used, also known as the Jaccard similarity coefficient, that is defined as:

$$J(A,B) = \frac{|A \cap B|}{|A \cup B|} \quad 0 \le J(A,B) \le 1 \quad \text{if } A, B \in \emptyset \Rightarrow J(A,B) = 1 \quad (1)$$

Table 1. Frequency of concepts from MeSH, Agrovoc and DrugBank identified in the LIVIVO corpus of 63 Million citations.

Dictionary	# of found concepts	# of Documents with concepts
MeSH	531,795,910	40,665,773
DrugBank	47,486,317	9,584,408
Agrovoc	447,766,801	39,947,272

The intersections, complements, and the Jaccard Index are calculated against each other for MeSH, Agrovoc and DrugBank. Initially, the complete sets of found concepts in the LIVIVO corpus are compared to each other. This comparison is

---

[1] http://www.livivo.de/ [Accessed October 2016].
[2] https://uima.apache.org/ [Accessed October 2016].
[3] https://www.nlm.nih.gov/mesh/ [Accessed October 2016].
[4] http://www.drugbank.ca/ [Accessed October 2016].
[5] http://aims.fao.org/standards/agrovoc/ [Accessed October 2016].
[6] https://datahub.io/dataset/livtdm [Accessed October 2016].

called `OverlapTotal`. The top-k most frequent concepts are compared in order to represent the requirement of information retrieval for accuracy and specificity. In information retrieval, Precision@k and Recall@k are calculated to evaluate relevancy performance. Appropriate numbers for k are 10, 20, 50, 100, 200, or 1000 [2]. In order to represent an overview of the total impact on relevancy ranking, k is set to 1000. The top 1000 most frequent concepts are taken for each of the ontologies. This comparison is called `OverlapTop1k`.

3   Results

In Table 2, the statistics for `OverlapTotal` and `OverlapTop1k` are shown. The results of set similarity measures for `MeSH`, `DrugBank`, and `Agrovoc` show that the pattern of `OverlapTotal` and of `OverlapTop1k` for each ontology differ markedly. In Fig. 1, the Venn Diagrams for `OverlapTotal` is shown as part A and the Venn Diagram for `OverlapTop1k` is shown as part B. For the total set, `MeSH` and `AGROVOC` show a relatively wide overlap with a Jaccard index of 0.0896,

Table 2. Set comparison of overlapping semantic entities between `MeSH`, `DrugBank`, and `AGROVOC`.

Dictionaries	$A \cap B$	$A \backslash B$	$B \backslash A$	$J(A, B)$
A = MeSH, B = DrugBank	909	26,272	998	0.0322
A = DrugBank, B = Agrovoc	162	1,745	26,937	0.0056
A = Agrovoc, B = MeSH	4,466	22,633	22,715	0.0896
A = MeSH Top1k, B = DrugBank Top1k	21	979	979	0.0106
A = DrugBank Top1k, B = Agrovoc Top1k	19	981	981	0.0095
A = Agrovoc Top1k, B = MeSH Top1k	473	527	527	0.3097

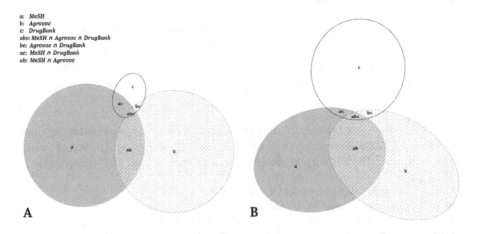

a: MeSH
b: Agrovoc
c: DrugBank
abc: MeSH ∩ Agrovoc ∩ DrugBank
bc: Agrovoc ∩ DrugBank
ac: MeSH ∩ DrugBank
ab: MeSH ∩ Agrovoc

A   **B**

Fig. 1. Venn Diagramm A shows the overlap of the complete sets of `MeSH`, `DrugBank`, and `AGROVOC`. Venn Diagramm B shows the overlap of the top 1000 concepts amongst the ontologies.

while DrugBank and MeSH overlap to a relatively small extent with 0.0322. The smallest overlap is between DrugBank and Agrovoc with an index of 0.0056. For OverlapTop1k, the overlaps change dramatically for MeSH and Agrovoc with a shift of the Jaccard index from 0.0896 to 0.3097. MeSH and DrugBank has a shift of the index to 0.0106. The results reflect the overall specificity of DrugBank concepts. MeSH and AGROVOC share concepts to a large extent, because both include generic concepts that are found at high frequency. The results indicate potential difficulties when employing ontologies in literature retrieval.

4   Conclusion

MeSH and AGROVOC contain overlapping, generic concepts in the LIVIVO corpus with a high frequency whereas DrugBank shows a high specificity. Generic overlaps perturb the specificity of relevant results. In order to over overcome this perturbation, either generic concepts can be mapped between the two ontologies, or generic concepts can completely be discard. The mapping of generic concepts links shared concepts without unnecessary noise. The removal of generic concepts results in an information loss. Nevertheless, the information loss does potentially still increase the retrieval performance. Future work will focus on the incorporation of real search queries in order to evaluate the potential solutions for filtering life science ontologies. Furthermore, additional biomedical ontologies will be incorporated for usage in literature retrieval.

References

1. Bornmann, L., Mutz, R.: Growth rates of modern science: a bibliometric analysis based on the number of publications and cited references. J. Assoc. Inf. Sci. Technol. **66**, 2215–2222 (2015)
2. Büttcher, S., Clarke, C.L.A., Cormack, G.V.: Information Retrieval: Implementing and Evaluating Search Engines. The MIT Press, Cambridge (2010)
3. Caracciolo, C., Stellato, A., Morshed, A., Johannsen, G., Rajbhandari, S., Jaques, Y., Keizer, J.: The AGROVOC linked dataset. Semant. Web **4**(3), 341–348 (2013)
4. Law, V., Knox, C., Djoumbou, Y., Jewison, T., Guo, A.C., Liu, Y., Maciejewski, A., Arndt, D., Wilson, M., Neveu, V., Tang, A., Gabriel, G., Ly, C., Adamjee, S., Dame, Z.T., Han, B., Zhou, Y., Wishart, D.S.: Drugbank 4.0: shedding new light on drug metabolism. Nucl. Acids Res. **42**(Database issue), D1091–D1097 (2014)
5. Müller, B., Hagelstein, A.: Beyond metadata enriching life science publications in livivo with semantic entities from the linked data cloud. In: Joint Proceedings of the Posters and Demos Track of the 12th International Conference on Semantic Systems - SEMANTiCS2016 and the 1st International Workshop on Semantic Change & Evolving Semantics (SuCCESS 2016), Co-located with the 12th International Conference on Semantic Systems (SEMANTiCS 2016), Leipzig, Germany (2016)
6. Rogers, F.B.: Medical subject headings. Bull. Med. Libr. Assoc. **51**, 114–116 (1963)
7. Tanenblatt, M., Coden, A., Sominsky, I.: The conceptmapper approach to named entity recognition. In: Calzolari, N., Choukri, K., Maegaard, B., Mariani, J., Odijk, J., Piperidis, S., Rosner, M., Tapias, D. (eds.) Proceedings of the Seventh International Conference on Language Resources and Evaluation (LREC'10). European Language Resources Association (ELRA), Valletta, Malta (2010)

Building Citation Networks with SPACIN

Silvio Peroni[1]([✉]), David Shotton[2], and Fabio Vitali[1]

[1] DASPLab, DISI, University of Bologna, Bologna, Italy
{silvio.peroni,fabio.vitali}@unibo.it
[2] Oxford e-Research Centre, University of Oxford, Oxford, UK
david.shotton@oerc.ox.ac.uk

Abstract. In this demo paper we introduce SPACIN, one of the main tools used in the OpenCitations Project for producing RDF-based citation data from information available in trusty sources, such as Europe PubMed Central, Crossref, and ORCID.

Keywords: OCC · OpenCitations · OpenCitations Corpus · SPAR Ontologies · Semantic publishing · Citation database

1 Introduction

The OpenCitations Project (http://opencitations.net) [3] has recently created a new instantiation of its citation database, with an integrated SPARQL endpoint[1] and a browsing interface to support data consumers [5]. This database, the OpenCitations Corpus (OCC), is an open repository of scholarly citation data made available under a Creative Commons public domain dedication (CC0)[2], which provides accurate bibliographic references harvested from the scholarly literature, that others may freely build upon, enhance and reuse for any purpose, without restriction under copyright or database law – e.g. Wikidata[3] has already started to use OCC for enriching the description of scholarly papers. The OCC is evolving dynamically in time, since new articles are continuously gathered (and then processed), and it enables the creation of incoming and outgoing links easily by means of the format used to store all the data, i.e. RDF.

In this demo paper we introduce the *SPAR Citation Indexer*, or *SPACIN*, which is the primary mechanisms used in OpenCitations for producing all the data included in the OCC. In particular, we show which kind of input SPACIN is able to process, what kinds of RDF data it produces, and what are the ontologies SPACIN uses for describing such data. All the code of OpenCitations, including SPACIN, is released with an ISC license[4] and, thus, can be freely reused in different context and according to different purposes.

RASH: https://w3id.org/oc/paper/spacin-demo-ekaw2016.html
[1] https://w3id.org/oc/sparql.
[2] https://creativecommons.org/publicdomain/zero/1.0/legalcode.
[3] http://wikidata.org/.
[4] https://opensource.org/licenses/ISC.

© Springer International Publishing AG 2017
P. Ciancarini et al. (Eds.): EKAW 2016 Satellite Events, LNAI 10180, pp. 162–166, 2017.
DOI: 10.1007/978-3-319-58694-6_23

2   The SPAR Citation Indexer

The *SPAR Citation Indexer*, a.k.a. *SPACIN*, is a script and a series of Python classes that allow one to process particular JSON files containing the bibliographic reference lists of papers, produced by another script included in the OpenCitations GitHub repository[5] that retrieves such lists by querying the Europe PubMed Central API[6]. An excerpt of an input JSON file used by SPACIN is introduced as follows:

```
{
    "localid": "MED-27193261", "doi": "10.1038/ncomms11627",
    "curator": "BEE EuropeanPubMedCentralProcessor",
    "source_provider": "Europe PubMed Central",
    "source": "http://www.ebi.ac.uk/europepmc/webservices/rest/PMC4874038/
        fullTextXML",
    "references": [ {
        "bibentry": "Weaver C. T, Elson C. O, Fouser L. A. & Kolls J. K. The Th17
            pathway and inflammatory diseases of the intestines, lungs, and skin
            . Annu. Rev. Pathol. 8, 477\u2013512 (2013). PMID: 23157335",
        "doi": "10.1146/annurev-pathol-011110-130318", ... }, ... ], ...
}
```

SPACIN processes such JSON files and retrieves additional metadata information about all the citing/cited articles by querying the Crossref API[7] and the ORCID API[8]. These API are also used to disambiguate bibliographic resources and agents by means of the identifiers retrieved (e.g., DOI, ISSN, ISBN, ORCID, URL, and Crossref member URL). Once SPACIN has retrieved all these metadata, RDF resources are created (or reused, if they have been already added in the past) and stored in the file system in JSON-LD format. In addition, they are also uploaded to the triplestore (via SPARQL UPDATE protocol) specified by the variable *triplestore_url* in the file *conf_spacin.py*.

SPACIN stores all the metadata relevant to bibliographic entities by using the OCC metadata model [4] summarised in Fig. 1. The ontological terms of such metadata model are collected within an ontology called the OpenCitations Ontology (OCO)[9], which includes several terms from the SPAR Ontologies[10] [2] and other vocabularies. In particular, the following six bibliographic entity types occur in the datasets created by SPACIN:

- **bibliographic resources** (br), class `fabio:Expression` – resources that either cite or are cited by other bibliographic resources (e.g. journal articles), or that contain such citing/cited resources (e.g. journals);
- **resource embodiments** (re), class `fabio:Manifestation` – details of the physical or digital forms in which the bibliographic resources are made available by their publishers;

---

[5] https://github.com/essepuntato/opencitations.

[6] https://europepmc.org/RestfulWebService.

[7] http://api.crossref.org/.

[8] http://members.orcid.org/api/.

[9] https://w3id.org/oc/ontology.

[10] http://www.sparontologies.net/.

- **bibliographic entries** (be), class `biro:BibliographicReference` – literal textual bibliographic entries occurring in the reference lists of bibliographic resources;
- **responsible agents** (ra), class `foaf:Agent` – names of agents having certain roles with respect to the bibliographic resources (i.e. names of authors, editors, publishers, etc.);
- **agent roles** (ar), class `pro:RoleInTime` – roles held by agents with respect to the bibliographic resources (e.g. author, editor, publisher);
- **identifiers** (id) (class `datacite:Identifier`) – external identifiers (e.g. DOI, ORCID, PubMedID) associated to bibliographic resources and agents.

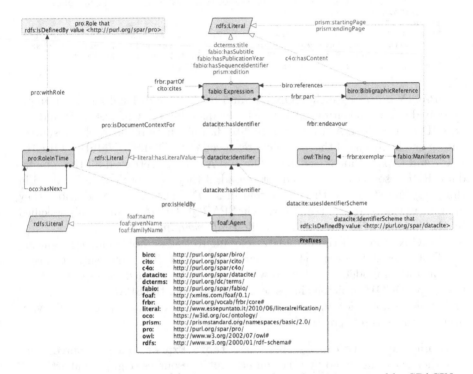

Fig. 1. The Graffoo diagram [1] of the main ontological entities created by SPACIN.

Upon initial curation by SPACIN, a URL is assigned to each entity, according to the following schema:

```
[corpus URL] + [sub-dataset ID] + / + [entity ID]
```

The `corpus URL` (which is https://w3id.org/oc/corpus/ for the OCC) identifies the entire dataset one is creating. It is possible to set it up by modifying the variable *base_iri* specified in the *conf_spacin.py* file in the GitHub repository. The `sub-dataset` ID is the two-letter short name for the class of items we are taking into consideration (e.g. "br" for bibliographic references), while the `entity ID` is a number that identifies a particular item that is unique among resources of the same type. For instance, the entity https://w3id.org/oc/corpus/br/525205 in the OCC identifies the 525,205[th] bibliographic resource stored by SPACIN – where the two-letter short name for the class of items plus "/" plus the number ("br/525205" in the example) is called *corpus identifier*, since it allows the unique identification of any entity within the dataset one is building by means of SPACIN.

Each of the entities created by SPACIN is associated with metadata describing its provenance using the PROV-O[11] ontology and its PROV-DC extension[12] (e.g. https://w3id.org/oc/corpus/br/525205/prov/se/1 in the OCC). In particular, SPACIN stores information about the curatorial activities related to each entity, the curatorial agents involved, and their roles.

All these RDF data are stored in BibJSON[13] encoded as JSON-LD, defined through an appropriate JSON-LD context[14] which hides the complexity of the model (shown in Fig. 1) behind natural language keywords. For instance, the following excerpt is the JSON-LD linearisation of the aforementioned "br/525205" entity:

```
{
  "iri": "gbr:525205", "a": [ "article", "document" ],
  "label": "bibliographic resource 525205 [br/525205]",
  "title": "The Electronic Patient Reported Outcome Tool: Testing Usability
      and Feasibility of a Mobile App and Portal to Support Care for Patients
      With Complex Chronic Disease and Disability in Primary Care Settings",
  "year": "2016", "part_of": "gbr:476045",    "format": "gre:217773",
  "reference": [ "gbe:727463", "gbe:727473", ... ],
  "citation": [ "gbr:1095446", "gbr:525190", ... ],
  "contributor": [ "gar:1591192", "gar:1591193", ... ],
  "identifier": [ "gid:816999", "gid:816997", ... ]
}
```

In this excerpt, "iri" defines the URL of the resource in consideration (where "gbr:" is a prefix for https://w3id.org/oc/corpus/br/), while a, label, title, year, part_of, reference, citation, format, contributor, and identifier stand for rdf:type, rdfs:label, dcterms:title, fabio:hasPublicationYear, frbr: partOf, biro:references, cito:cites, frbr:embodiment, c4o: isDocumentCon textFor, and datacite:hasIdentifier respectively (where "gbe:" is a prefix for https://w3id.org/oc/corpus/be/, "gre:" for https://w3id. org/oc/corpus/re/, and "gid:" for https://w3id.org/oc/corpus/id/).

Additional information about how SPACIN stores citation data, and the way they are represented in RDF, are detailed in the OCC Metadata Document [4].

3   Conclusions

In this paper we have introduced SPACIN, the SPAR Citation Indexer, that is a tool for creating RDF-based citation data starting from information made available by trusty sources, such as Europe PubMed Central, Crossref, and ORCID.

---

[11] https://www.w3.org/TR/prov-o/.

[12] https://www.w3.org/TR/prov-dc/.

[13] http://okfnlabs.org/bibjson/.

[14] https://w3id.org/oc/corpus/context.json.

In the future, we plan to extend SPACIN for enabling the parallel execution of multiple instances of such script, so as to increase the number of new bibliographic entities created by SPACIN each day – which currently is around 20,000 new citing/cited bibliographic resources, according to the statistics related with the OCC.

References

1. Falco, R., Gangemi, A., Peroni, S., Shotton, D., Vitali, F.: Modelling OWL ontologies with graffoo. In: Presutti, V., Blomqvist, E., Troncy, R., Sack, H., Papadakis, I., Tordai, A. (eds.) ESWC 2014. LNCS, vol. 8798, pp. 320–325. Springer, Cham (2014). doi:10.1007/978-3-319-11955-7_42
2. Peroni, S.: The semantic publishing and referencing ontologies. In: Peroni, S. (ed.) Semantic Web Technologies and Legal Scholarly Publishing. Law, Governance and Technology Series, pp. 121–193. Springer, Heidelberg (2014). http://dx.doi.org/10.1007/978-3-319-04777-5_5
3. Peroni, S., Dutton, A., Gray, T., Shotton, D.: Setting our bibliographic references free: towards open citation data. J. Doc. **71**(2), 253–277 (2015). http://dx.doi.org/10.1108/JD-12-2013-0166
4. Peroni, S., Shotton, D.: Metadata for the OpenCitations Corpus. Figshare (2016). https://dx.doi.org/10.6084/m9.figshare.3443876
5. Peroni, S., Shotton, D., Vitali, F.: Freedom for bibliographic references: opencitations arise. In: Proceedings of 2016 International Workshop on Linked Data for Information Extraction (LD4IE 2016) (2016). https://w3id.org/oc/paper/occ-lisc2016.html

Graph-Based Relation Validation Method

Rashedur Rahman[1]($\boxtimes$), Brigitte Grau[2], and Sophie Rosset[3]

[1] IRT SystemX, LIMSI, CNRS, Université Paris-Saclay, Orsay, France
rashedur.rahman@irt-systemx.fr
[2] LIMSI, CNRS, ENSIIE, Université Paris-Saclay, Orsay, France
brigitte.grau@limsi.fr
[3] LIMSI, CNRS, Université Paris-Saclay, Orsay, France
sophie.rosset@limsi.fr

Abstract. In this paper we present a relation validation method for KBP slot filling task by exploring some graph features to classify the candidate slot fillers as correct or incorrect. The proposed features with voting feature collectively performs better than the baseline voting feature.

1 Introduction

Relation extraction and validation plays an important role in information extraction task like slot filling (SF) for knowledge base population (KBP). SF defines the task of finding the filler-entity (or entity) from texts by justifying the relation (or slot) of a given entity (the query). It requires entity level relation extraction based on the mention level relations (MLR) and entity linking that are dependent on each other. Traditional MLR extractor does not achieve satisfying precision and recall for SF task [1] and it often results high confidence score for incorrect relations because of limited features and training data. We propose to work at the entity level for validating relations because we can use additional features that cannot be used at mention level.

In this paper we explore community-graph based features for validating SF relations that were not explored before. The community graph is made of neighbor entities. Han et al. [2] proposed *referent graph* for collective entity linking where they took into account the semantic relations among the neighbor entities. Friedl et al. [3] discussed the use of different centrality measurements to find the important and influential nodes in social networks. Solá et al. [4] explored the concepts of eigenvector centrality in multiplex networks and showed the existence of such centrality which is unique. Information theoretic measurements have also been proposed for knowledge discovery in complex networks [5], or for validating answer in question-answering systems [6]. We propose to explore community network and information theoretic concepts for validating relations between query and candidate entities. We consider relation validation as a binary classification task where the candidate filler entities are generated by MLR extractor. Here we do not evaluate the KBP SF task but evaluate relation validation. We observe the classification performances of different feature sets and show that

© Springer International Publishing AG 2017
P. Ciancarini et al. (Eds.): EKAW 2016 Satellite Events, LNAI 10180, pp. 167–171, 2017.
DOI: 10.1007/978-3-319-58694-6_24

the model including all features increases the F-score by 2.6% compare to the baseline voting system.

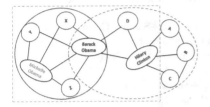

Fig. 1. Community graph (Color figure online)   **Fig. 2.** Knowledge graph

2   Method Description

Let, a graph $G = (V, E)$, query relation (slot) R_q, query entity $v_q \epsilon V$, candidate filler-entities $V_c = \{v_{c1}, v_{c2}, \ldots, v_{cn}\} \epsilon V$ where $R_q = e(v_q, v_c) \epsilon E$. The candidate list is generated by relation extractor. Suppose other semantic relations $R_o \epsilon E$ where $R_o \neq R_q$. We define the task to classify whether a filler-entity c of C_v is correct or incorrect for a query relation (R_q) by analyzing the communities of query entity and candidate fillers. Figure 1 shows an example of community based relation validation task where the query entity, type and slot name are *Barack Obama, person* and *spouse* accordingly. The slot filler candidates are *Michelle Robinson* and *Hilary Clinton* that are linked to *Barack Obama* by *spouse* relation hypothesis. The communities of *Barack Obama* (green rectangle), *Michelle Robinson* (purple circle) and *Hilary Clinton* (orange ellipse) are constructed by *in_same_sentence* relation which means the pair of entities are mentioned in the same sentences in the texts. We want to classify *Michelle Robinson* as the correct slot filler based on community analysis. We create the community graph (CG) of entities from the knowledge graph (KG) as illustrated in Fig. 2. The KG is generated by applying some processing on the texts that includes named entity recognition (NER), sentence splitting, relation extraction (RE). The RE extracts the semantic relation between a pair of entities and gives a confidence score. The KG represents the documents, sentences, mentions as nodes and their relations. Entity mentions are connected to entities in the CG and an edge connects a pair of entities in the CG based on the associated relation hypothesis. Since the CG is constructed based on the KG, the semantics is maintained in the CG. We include *person, location and organization* typed entities as the community members in our community-graph-based analysis. A community is built with a set of entities which are mostly inter-related.

We assume that a correct filler-entity of a SF query should be a strong member in the community of the query entity and such community can be extracted from the texts by extracting semantic relations and/or based on their existences in the same sentences. We hypothesize that the *network density* (Eq. 1) of a community of a correct filler-entity with the query entity should be higher than a community of an incorrect filler-entity with the query entity. In Fig. 1 the community of *Michelle Robinson* with *Barack Obama* is more dense than the community of *Hilary Clinton*. Eigenvector centrality [7] measures the influence of a neighbor node to measure the centrality of a node in a community. We quantify the influence of the candidate fillers in the community of a query entity by calculating the absolute difference between the eigenvector centrality scores of the query and a filler entity. We hypothesize that the difference should be smaller for a correct filler than an incorrect filler. We also hypothesize that the *mutual information* (Eq. 3) and *similarity* (Eq. 2) between the community of a correct filler and the community of the query entity should be higher than an incorrect filler. The community of an entity (query entity or a candidate filler-entity) is expanded up to level 3 to measure the eigenvector centrality and mutual information. Additionally, we defined 6 ratios at the collection level as features based on the Eqs. 4 to 9. For example, we calculate the ratio of occurrences of a filler-entity mention that are associated by the query relation with the query entity to the total number of mentions of that filler (Eq. 4). We also include the confidence score (given by a relation extractor) as feature.

$$\rho_{network} = \frac{number\,of\,existing\,edges}{number\,of\,possible\,edges} \tag{1}$$

$$cosine\,similarity = \frac{|X \cap Y|}{\sqrt{|X||Y|}} \tag{2}$$

where, X and Y are the set of community-members of query and filler entity accordingly

$$MI(X,Y) = H(X) + H(Y) - H(X,Y) \tag{3}$$

where, $H(X) = -\sum_{i=1}^{n} p(x_i)\,log_2(p(x_i))$, X and Y are the communities of query and filler entity accordingly and p(x) refers to the probability of centrality degree of a community-member

$$r_{mention}(e_c) = \frac{\#\,of\,mentions\,associated\,with\,e_q}{total\,number\,of\,mentions} \tag{4}$$

$$r_{hyp1}(e_q, e_c) = \frac{\#\,of\,documents\,with\,r_q(e_q, e_c)\,hyp.}{total\,number\,of\,r_q(e_q, e_c)\,hyp.} \tag{5}$$

$$r_{hyp2}(e_q, e_c) = \frac{total\,number\,of\,r_q(e_q, e_c)\,hyp.}{\#\,of\,sentences\,containing\,e_q\,and\,e_c} \tag{6}$$

$$r_{doc}(e_q, e_c) = \frac{\#\,of\,documents\,with\,r_q(e_q, e_c)\,hyp.}{total\,document\,count\,with\,all\,candidates} \tag{7}$$

$$relFreq_{hyp}(e_c) = \frac{number\ of\ r_q(e_q, e_c)\ hyp.}{total\ r_q\ count\ with\ all\ candidates} \qquad (8)$$

$$r_{entity}(r_q, e_q, e_c) = \frac{\#\ of\ different\ entities\ in\ all\ r_q(e_q, e_c)\ sentences}{total\ entity\ count\ in\ all\ r_q(e_q, e_c)\ sentences} \qquad (9)$$

3   Dataset and Experiments

We use the assessments of 2014 Cold Start Slot Filling (CSSF) evaluation task that contains 100 queries (50 for PERSON and 50 for ORGANIZATION) for building our corpus. Each PERSON query includes 16 slots (*per:spouse*, *per:children* etc.) and an ORGANIZATION query contains 25 slots. The assessment files provide the correct and incorrect responses of the queries with the document reference that support the relation. We compile a subset of KBP-2014 evaluation corpus and we select 1942 documents for 7 slots (*org:top_members_employees*, *org:founded_by*, *per:statesorprovinces_of_residence*, *per:cities_of_residence*, *per:member_employees_of*, *per:children*, *per:spouse*) from the source document ids given in the assessment files and the queries because our current system limits to extract these relations. The documents (of correct and incorrect responses) are taken for a query slot if at least one responded slot filler string is justified as correct by NIST. The dataset includes 168 correct and 289 incorrect fillers from 97 queries. There exist multiple correct fillers for some queries because all the slots that we discuss here are multi-valued slots.

Table 1. Classification performances (in %)

Feature set	Precision	Recall	F-score
Voting (baseline)	76.0	72.9	72.6
Graph + voting	75.0	73.5	73.5
Graph + collective (all) + voting	75.4	74.5	74.6
Graph + collective (Eqs. 5 and 6) + voting	**76.0**	**75.2**	**75.2**

We trained several binary classifiers in Weka3.8 evaluated with 10 fold cross validation and the best F-score was achieved by SMO classifier. We group the relation validation features into 3 sets: (i) *graph features*: Eqs. 1–3 and eigenvector centrality. (ii) *collective*: Eqs. 4 to 9 and confidence score (iii) *voting (baseline)*: counts the maximum vote of a filler. Table 1 depicts the classification performances of different feature sets. The voting baseline obtains an F-score of 72.6%. The graph features and voting collectively obtain an F-score of 73.5%. We achieve the highest F-score of 75.2 (2.6% higher than voting baseline) by using all the graph features, two of the collective features (Eqs. 4 and 5) and the voting features. The best scoring features have been selected by using attribute selection method (ReliefF) in Weka. We evaluate the proposed features for validating relation between a pair of entities instead of evaluating its impact on the SF task because it has some requirements that we do not consider here. However, the experimental results strongly support the proposed community-graph based features for validating relations that could be very effective for SF tasks.

4 Conclusion

In this paper we explored some community-graph based and corpus level collective features for validating relations that obtained promising results to classify correct and incorrect relations. The proposed graph features increased the F-score by 2.6% that strongly argues to continue graph based analysis for validating relation hypothesis.

References

1. Surdeanu, M., Ji, H.: Overview of the English slot filling track at the TAC 2014 knowledge base population evaluation. In: Proceedings of the Text Analysis Conference (TAC 2014) (2014)
2. Han, X., Sun, L., Zhao, J.: Collective entity linking in web text: a graph-based method. In: Proceedings of the 34th international ACM SIGIR conference on Research and development in Information Retrieval, pp. 765–774. ACM (2011)
3. Friedl, D.M.B., Heidemann, J., et al.: A critical review of centrality measures in social networks. Bus. Inf. Syst. Eng. 2(6), 371–385 (2010)
4. Solá, L., Romance, M., Criado, R., Flores, J., del Amo, A.G., Boccaletti, S.: Eigenvector centrality of nodes in multiplex networks. Chaos Interdiscip. J. Nonlinear Sci. 23(3), 033131 (2013)
5. Holzinger, A., Ofner, B., Stocker, C., Calero Valdez, A., Schaar, A.K., Ziefle, M., Dehmer, M.: On graph entropy measures for knowledge discovery from publication network data. In: Cuzzocrea, A., Kittl, C., Simos, D.E., Weippl, E., Xu, L. (eds.) CD-ARES 2013. LNCS, vol. 8127, pp. 354–362. Springer, Heidelberg (2013). doi:10. 1007/978-3-642-40511-2_25
6. Magnini, B., Negri, M., Prevete, R., Tanev, H.: Is it the right answer? Exploiting web redundancy for answer validation. In: Proceedings of the 40th Annual Meeting on Association for Computational Linguistics, pp. 425–432. Association for Computational Linguistics (2002)
7. Bonacich, P., Lloyd, P.: Eigenvector-like measures of centrality for asymmetric relations. Soc. Netw. 23(3), 191–201 (2001)

Probabilistic Inductive Logic Programming on the Web

Fabrizio Riguzzi[1](✉), Riccardo Zese[2], and Giuseppe Cota[2]

[1] Dipartimento di Matematica e Informatica, University of Ferrara,
Via Saragat 1, 44122 Ferrara, Italy
`fabrizio.riguzzi@unife.it`
[2] Dipartimento di Ingegneria, University of Ferrara,
Via Saragat 1, 44122 Ferrara, Italy
`{riccardo.zese,giuseppe.cota}@unife.it`

Abstract. Probabilistic Inductive Logic Programming (PILP) is gaining attention for its capability of modeling complex domains containing uncertain relationships among entities. Among PILP systems, `cplint` provides inference and learning algorithms competitive with the state of the art. Besides parameter learning, `cplint` provides one of the few structure learning algorithms for PLP, SLIPCOVER. Moreover, an online version was recently developed, `cplint` on SWISH, that allows users to experiment with the system using just a web browser. In this demo we illustrate `cplint` on SWISH concentrating on structure learning with SLIPCOVER. `cplint` on SWISH also includes many examples and a step-by-step tutorial.

1   Introduction

Probabilistic Inductive Logic Programming (PILP) [3] uses Probabilistic Logic Programming (PLP) for modeling in domain characterized by uncertain relationships among entities.

One of most successful approaches to PLP is based on the distribution semantics [8] where a probabilistic program defines a probability distribution over non probabilistic programs, called worlds. The probability of a query is simply the sum of the probability of the worlds where the query is true. Various languages follow the distribution semantics such as Probabilistic Logic Programs, Logic Programs with Annotated Disjunctions (LPADs), CP-logic and ProbLog.

Many systems for performing inference and learning with these languages have been proposed in the past 20 years. Among them, `cplint` provides an interesting mix of algorithms, including structure learning algorithms.

`cplint` on SWISH [6] is a web application for running `cplint` with just a web browser: the algorithms run on a server and the user can post queries and see the results in his browser. `cplint` on SWISH is available at http://cplint. lamping.unife.it and Fig. 1 shows its interface.

`cplint` on SWISH is based on SWISH, a web framework for Logic Programming using features and packages of SWI-Prolog and its Pengines library.

P. Ciancarini et al. (Eds.): EKAW 2016 Satellite Events, LNAI 10180, pp. 172–175, 2017.
DOI: 10.1007/978-3-319-58694-6_25

Fig. 1. cplint on SWISH interface.

SWISH allows the user to write a Logic Program in a browser window and ask a query over it. The query and the program are sent to a server using JavaScript. The server then builds a Pengine (Prolog Engine) that evaluates the query and returns answers for it to the user. Both the web server and the inference back-end are run entirely within SWI-Prolog.

cplint on SWISH uses the language of LPADs and includes two inference algorithms: PITA, that uses knowledge compilation, and MCINTYRE, that uses Monte Carlo sampling. For parameter learning EMBLEM [1] is available while SLIPCOVER [2] can be used for structure learning.

cplint on SWISH is similar to ProbLog2 [4] that has also an online version[1]. Problog2 offer inference and learning algorithms for ProbLog. In the online version, users are allowed to write programs and run algorithms on a server with a browser. ProbLog2 is written in Python, runs in an Python HTTP server and exploits the ACE editor[2] which is written in JavaScript.

The main difference between cplint on SWISH and ProbLog2 is that the first offers also structure learning. Moreover, cplint on SWISH uses a Prolog-only software stack. ProbLog2, instead, relies on several different technologies, including JavaScript, Python 3 and the DSHARP compiler. In particular, it writes intermediate files to disk in order to call external programs such as DSHARP while we work in main memory only.

With both cplint on SWISH and ProbLog2 users who want to experiment with PILP can do it without the need to install a system, a procedure which is often complex, error prone and limited mainly to the Linux platform. However, since it is impossible to predict the load of the server, the system is more targeted at development, while for production it is recommended to use the standalone version of cplint.

One of the main objectives of cplint on SWISH is to reach out to a wider audience and popularize PILP, similarly to what is done for the functional probabilistic language Church [5], which is equipped with the webchurch system for compiling Church programs into JavaScript. To try to achieve this goal, cplint

---

[1] https://dtai.cs.kuleuven.be/problog/.

[2] https://ace.c9.io/.

on SWISH includes various learning examples[3]: Machines (shown below), Registration, Bongard, Shop, Hidden Markov Model, Mutagenesis, University. Moreover, a complete online tutorial is available [7] at http://ds.ing.unife.it/~gcota/plptutorial/.

2 EMBLEM and SLIPCOVER

EMBLEM and SLIPCOVER in `cplint` on SWISH take as input a program divided in five parts: (1) a preamble where all the parameters (such as the maximum number of iterations and the verbosity level) are set, (2) a background knowledge that contains information valid for all interpretations, (3) an initial LPAD if there is one, (4) a language bias for guiding the learning phase. In particular, for EMBLEM it contains the declarations of the input and output predicates. Basically, input predicates are those for which we do not want to learn parameters, while we want to learn parameters for output predicates. SLIPCOVER uses also bias declaration in the style of Progol and Aleph: atoms for `modeh/2` specify the literals that can appear in the head of clauses while atoms for `modeb/2` specify the atoms that can appear in the body of clauses. Moreover, SLIPCOVER requires the use of the `determination/2` predicate, as in Aleph, for specifying which predicates can appear in the body of clauses.

Using the machine dataset of the ACE data mining system[4] as a running example[5] we have:

```
modeh(*,class(sendback)).
modeb(*,not_replaceable(-comp)).
modeb(*,replaceable(-comp)).
determination(class/1,replaceable/1).
determination(class/1,not_replaceable/1).
```

Finally, (5) the last part contains example interpretations. Here, we can use two different representations shown below, *models* or *keys*, as in ACE. The first specifies an example interpretation as a list of Prolog facts surrounded by `begin(model(<name>))` and `end(model(<name>))`. In the latter the facts can be directly listed using the first argument as the example name.

```
begin(model(1)).              class(1,sendback).
class(sendback).              neg(1,class(fix)).
neg(class(fix)).              worn(1,engine).
worn(engine).
end(model(1)).
```

We can also define folds and which examples are included so that learning can be performed by just asking the query `induce(<folds>,P)`. The learned program will be returned in variable P.

[3] http://cplint.lamping.unife.it/example/learning/learning_examples.swinb.
[4] https://dtai.cs.kuleuven.be/ACE/.
[5] http://cplint.lamping.unife.it/example/learning/mach.pl.

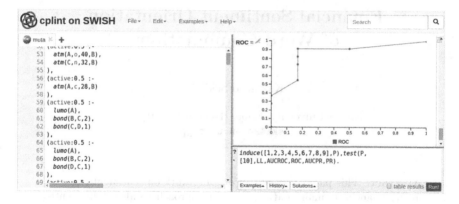

Fig. 2. ROC curve for the mutagenesis dataset.

cplint on SWISH has also facilities for testing the learned models: predicate test/7 takes as input the learned program and the testing folds and returns the log-likelihood, the areas under the precision-recall and receiver operating characteristics curves together with the set of points forming the curves themselves. These set of points can also be drawn on the screen as the respective curves, as shown in Fig. 2. cplint on SWISH offers also a separate AUC calculator[6] that takes as input the list of examples together with their sign and probability.

Acknowledgement. This work was supported by the "GNCS-INdAM".

References

1. Bellodi, E., Riguzzi, F.: Expectation maximization over binary decision diagrams for probabilistic logic programs. Intell. Data Anal. **17**(2), 343–363 (2013)
2. Bellodi, E., Riguzzi, F.: Structure learning of probabilistic logic programs by searching the clause space. Theor. Pract. Log. Program. **15**(2), 169–212 (2015)
3. Raedt, L., Kersting, K.: Probabilistic inductive logic programming. In: Raedt, L., Frasconi, P., Kersting, K., Muggleton, S. (eds.) Probabilistic Inductive Logic Programming. LNCS, vol. 4911, pp. 1–27. Springer, Heidelberg (2008). doi:10.1007/978-3-540-78652-8_1
4. Fierens, D., den Broeck, G.V., Renkens, J., Shterionov, D.S., Gutmann, B., Thon, I., Janssens, G., De Raedt, L.: Inference and learning in probabilistic logic programs using weighted boolean formulas. Theor. Pract. Log. Program. **15**(3), 358–401 (2015)
5. Goodman, N.D., Tenenbaum, J.B.: Probabilistic Models of Cognition. http://probmods.org
6. Riguzzi, F., Bellodi, E., Lamma, E., Zese, R., Cota, G.: Probabilistic logic programming on the web. Softw. Pract. Exp. **46**(10), 1381–1396 (2016)
7. Riguzzi, F., Cota, G.: Probabilistic logic programming tutorial. Assoc. Log. Program. Newsl. **29**(1) (2016). http://www.cs.nmsu.edu/ALP/2016/03/probabilistic-logic-programming-tutorial/
8. Sato, T.: A statistical learning method for logic programs with distribution semantics. In: Sterling, L. (ed.) ICLP-95, pp. 715–729. MIT Press, Cambridge (1995)

---

[6] http://cplint.lamping.unife.it/example/learning/exauc.pl.

Financial Sentiment Orientation
of Word Combinations

Kazuhiro Seki[✉]

Konan University, Kobe, Hyogo 658-8501, Japan
seki@konan-u.ac.jp

Abstract. This paper presents an ongoing work on sentiment analysis in the financial domain and explores an approach to identifying sentiment orientations of words for a given financial index. The proposed approach takes advantage of the movement of the given financial index and employs an information theoretic measure for estimating sentiment orientation of word combinations in an efficient way. Results on preliminary experiments are reported.

Keywords: Sentiment analysis · Text mining · Finance

1 Introduction

Reflecting the increasing volume of consumer generated media, such as blogs, micro blogs, and customer reviews, sentiment analysis for textual data has been actively studied for over a decade [3]. Sentiment analysis generally aims at classifying the polarity (i.e., positive or negative) or more specific emotional states (e.g., angry, sad, and happy) of an input sentence or document.

While sentiment analysis is generally used for judging sentiment of text toward its subject matter, it has been also applied to judging stock market sentiment. For example, Bollen et al. [2] analyzed Twitter data and showed that, when aggregated, they help to predict stock prices and volatility.

The present work also aims at the financial domain and attempts to devise an approach to estimating the polarized sentiment orientation (i.e., the degree of positivity and negativity) of a given word combination. Here, "positive" words are those indicative of an upward trend of a given financial index (e.g., Dow Jones Average) and "negative" words are those indicative of its downward trend. Such financial sentiment orientation scores can be a valuable in various tasks for financial text mining.

This paper reports my preliminary work on this subject. The proposed approach takes advantage of target index movements and news headlines in the same period to identify positive/negative news headlines and utilizes mutual information-based method. Also, the proposed approach is designed to be able to assess sentiment positivity/negativity of not only single words but also combinations of any number of words in a computationally efficient manner. The motivation of considering word combinations comes from an observation that

P. Ciancarini et al. (Eds.): EKAW 2016 Satellite Events, LNAI 10180, pp. 176–179, 2017.
DOI: 10.1007/978-3-319-58694-6_26

the polarity of single word cannot be often determined without looking at their co-occurring words (e.g., "increase" can be positive or negative depending on *what* increases). The proposed approach is tested on Thomson Reuters' Machine Readable News with stock prices being as the target indices.

2 Financial Sentiment Orientation

The proposed approach is composed of two steps: (1) aligning the past news events and the index movements to determine "positive" and "negative" headlines with regard to the index, and (2) computing sentiment orientation for each word or word combination appearing in the headlines. Each step is described in the next subsections.

2.1 Aligning Headlines and Index Movements

News headlines have date/time information indicating when they were transmitted, and the historical data of the target index are also abundant. By aligning the two data, news events having possibly caused the index movements, either positively or negatively, could be identified.

For the identification of news headlines, we just need to set two parameters, one indicating the maximum time span (window size), denoted as max_t, between the transmission of a news event and the actual movement of the target index. Another is the minimum percent change, denoted as min_c, big enough to be considered as a movement of the target index.

2.2 Computing Financial Sentiment Orientation

Following the work by Turney [4], this study defines the sentiment orientation (SO) of a word or word combination c as a log-odds ratio based on pointwise mutual information (PMI), i.e., $SO(c) = PMI_{pos}(c) - PMI_{neg}(c)$. Here, $PMI_{pos}(c)$ is defined as $\log_2(p_{pos}(c)/(p(c)N_{pos}/N))$ and $PMI_{neg}(c)$ as $\log_2(p_{neg}(c)/(p(c)N_{neg}/N))$, where $p(c)$, $p_{pos}(c)$, N_{pos}, and N denote the proportion of headlines containing c, the proportion of positive headlines containing c, and the number of positive headlines, and the total number of headlines, respectively ($p_{neg}(c)$ and N_{neg} are similarly defined). Plugging $PMI_{pos}(c)$ and $PMI_{neg}(c)$ in $SO(c)$ yields

$$SO(c) = \log_2 \frac{p_{pos}(c)}{p_{neg}(c)} \left(\frac{N_{neg}}{N_{pos}} \right)^2 . \tag{1}$$

As word combination c, we consider every possible combination appearing in input headlines. However, computing SO scores for such c leads to the combinatorial explosion and is computationally prohibitive. To mitigate the issue, a minimum relative frequency (min_{rf}) is set for efficient computation, which resembles the notion of the minimum support of the Apriori algorithm [1] as

---

Algorithm 1. Calculation of sentiment orientation for word combinations

---

Input: H_{pos}: a set of positive headlines
     H_{neg}: a set of negative headlines
     n: maximum length of a word combination
     min_{rf}: minimum relative frequency
Output: SO(c): sentiment orientation of every word combination w of length up to n
1:   $V \leftarrow$ vocabularies (words) in H_{pos} and H_{neg}
2:   $C_1 \leftarrow \{c \mid c \in V \wedge Cnt(c) > min_{rf}\}$
3:   **for** $k = 2$ **to** n **do**
4:      $C_k \leftarrow \{c_i \cup c_j \mid c_i, c_j \in C_{k-1} \wedge |c_i \cup c_j| = k \wedge Cnt(c_i \cup c_j) > min_{rf}\}$
5:   **end for**
6:   **for each** c in $\bigcup_{k=1}^{n} C_k$ **do**
7:      SO(c) $\leftarrow$ sentiment orientation of c
8:   **end for**

---

illustrated in Algorithm 1. In the algorithm, by only considering word combinations whose relative frequency is greater than min_{rf} (the 2nd and 4th lines in Algorithm 1), C_k (a set of word combinations of length k, $1 \leq k \leq n$) is kept manageable in size.

3   Evaluation

As news headlines and the target indices, the Machine Readable News (MRN) by Thomson Reuters from January 2013 to June 2014 and stock prices of the first section of the Tokyo Stock Exchange in the same period were utilized.

Using the data, positive/negative headlines were first identified as described in Sect. 2.1. With different financial indices, one can obtain different sets of positive/negative headlines. As an example, Table 1 shows some of the positive headlines identified based on stock price movements of Toyota in the order of absolute percent change of stock prices, where max_t and min_c were experimentally set to 15 min and +1.0%, respectively. Note that the original headlines were in Japanese and the following tables show English word-for-word translation to give a rough idea of what the headlines are about.

From Table 1, we can see that the most identified headlines are clearly positive for Toyota. On the other hand, identified negative headlines (not shown) were found to be mixed, containing both positive and negative headlines. Ideally, the sets of positive/negative headlines should be further separated to improve the purity, which is left for future work.

Table 1. Identified positive headlines.

Price	Mins	% change	Headlines
5570	10	3.7	US auto sales month percent increase pickup truck strong
4650	13	1.8	Month Indonesia car sales year-on-year percent increase
4260	13	1.5	Toyota month China auto sales surprised recovery

Next, using the identified headlines with presumed polarities, sentiment orientation of word combinations was calculated. First, positive/negative headlines for all brands were used in calculation. Although not presented here due to the space limitation, the word combinations with high SO scores were found to be mostly positive expressions in general. It should be mentioned that there were no negative word combinations (having negative SO scores). This is presumably due to the fact that automatically identified negative headlines were mixed with positive headlines, which made it difficult to spot frequent word combinations in the identified pseudo negative headlines.

Then positive/negative headlines for Toyota *only* were used for calculating SO scores. The expectation was that word combinations with high SO scores would be more brand-specific. Table 2 shows the some of the results.

Table 2. Positive words for Toyota.

Word combination c	SO(c)
Percent annual rate month US auto sales year-on-year	4.533
Favorability preceding buying major export shares	4.143
US	4.143
Weaker yen	1.488
Yen recovery	1.099

As expected, word combinations c with high SO scores reveal the characteristics of Toyota. Specifically, they reflect the fact that north America is the biggest market for Toyota and that the weak yen is favorable for the company. On the contrary, the negative word combinations (not shown) did not have clearly negative sentiment. Again, the result appears to be due to the low quality of automatically identified negative headlines used for this experiment.

Future work includes improving the quality of the negatives and evaluating the utility of the identified word combinations with sentiment orientations.

Acknowledgment. This work is partially supported by MEXT, Japan.

References

1. Agrawal, R., Srikant, R.: Fast algorithms for mining association rules in large databases. In: Proceedings of the 20th VLDB, pp. 487–499 (1994)
2. Bollen, J., Mao, H., Zeng, X.: Twitter mood predicts the stock market. J. Comput. Sci. **2**(1), 1–8 (2011)
3. Liu, B.: Sentiment Analysis: Mining Opinions, Sentiments, and Emotions. Cambridge University Press, Cambridge (2015)
4. Turney, P.D.: Thumbs up or thumbs down? Semantic orientation applied to unsupervised classification of reviews. In: Proceedings of the 40th ACL, pp. 417–424 (2002)

Towards Mining Patterns for Exploratory Search with *Keval* Algorithm

Tomasz Sosnowski and Jedrzej Potoniec[(✉)]

Faculty of Computing, Poznan University of Technology,
ul. Piotrowo 3, 60-965 Poznan, Poland
Tomasz.Sosnowski@student.put.poznan.pl, jpotoniec@cs.put.poznan.pl

Abstract. For a given set of URIs, finding their common graph patterns may provide useful knowledge. We present an algorithm searching for the best patterns while trying to extend the set of relevant URIs. It involves interaction with the user in order to supervise extension of the set.

1   Introduction

Finding graph patterns generalizing over a given set of Uniform Resource Identifiers (URIs) may be useful e.g. in an exploratory search task, where the user barely knows what is in an RDF graph and wants to explore it. The ability to extend a given set of URIs with supposedly best matching candidates may also be useful in this task. We propose *Keval*, an interactive algorithm for mining graph patterns from an initial set of URIs by interactive asking the user to classify selected URIs. The algorithm takes into account an RDF graph describing the URIs, but interacts with it only via a SPARQL endpoint. While it is possible to get all common triple patterns for a set of URIs by a simple SPARQL query, such a query is very sensitive to noisy data and not necessarily helpful for finding complex graph patterns, thus *Keval* may be useful as a robust heuristics for finding graph patterns, which is able to tolerate some errors in URI classification.

Starting from a relatively small initial set *Keval* iteratively extends the scope of search for URIs that may be interesting to the user. The mined patterns are expressed as SPARQL Basic Graph Patterns (BGPs) [12] consisting of only triple patterns (i.e. without filter expressions) and such that there is a single variable in every triple pattern.

In the paper, we use a prefix `dbr:` for a namespace http://dbpedia.org/resource/, `dbo:` for http://dbpedia.org/ontology/, `dbc:` for http://dbpedia.org/resource/Category: and `dct:` for http://purl.org/dc/terms/.

2   Related Work

Exploratory search, originating in information systems and the Web [7], was naturally adapted by the Semantic Web. Approaches proposed so far fall into one of the few categories: faceted browsing, visual interfaces and recommendations.

P. Ciancarini et al. (Eds.): EKAW 2016 Satellite Events, LNAI 10180, pp. 180–183, 2017.
DOI: 10.1007/978-3-319-58694-6_27

In a faceted browser, the user is presented with a dynamically generated interface based on the content of the underlying RDF graph [9]. A faceted browser may be also tailored to some specific type of RDF data, like in [6] or combined with natural language generation [2]. The user may also use a visual interface instead of typing the query, e.g. [13] allows for navigation over a displayed part of an RDF graph. Finally, a system may recommend vocabulary to the user during typing the query, like in [1]. The exploration may also happen after the results of the query are delivered, e.g. by organizing them into clusters [3].

Learning SPARQL BGPs was also considered in contexts other than exploratory search, e.g. for constructing features for classification in a supervised [4] or unsupervised [10] way.

3   Algorithm Description

At the beginning, the user provides two sets of URIs: relevant (URIs that are interesting for her) and irrelevant. Then the algorithm generates an initial set of graph patterns, based on the RDF triples containing relevant URIs. For every such a triple a new triple pattern with a variable in the place of the requested URI is generated. To avoid too specific or too general patterns, only these matching more than one, but no more than a fixed number of URIs, are retained. Then the remaining triple patterns are transformed into graph patterns.

Consider the following example concerning *the Apollo program*. The user specifies an address of a SPARQL endpoint containing *DBpedia* [5], to be used as the RDF graph. The user also provides two relevant URIs: dbr:Neil_Armstrong and dbr:Buzz_Aldrin, referring to two of the astronauts participating in the program. Now, the algorithm downloads all the triples containing either of the URIs in any position. The triples are then transformed into triple patterns, e.g. a triple dbr:Neil_Armstrong dct:subject dbc:Apollo_11 into ?uri dct:subject dbc:Apollo_11. This pattern is retained, because it matches also dbr:Buzz_Aldrin, but the pattern ?uri dct:subject dbc:X-15_program is removed, as Buzz Aldrin did not participate in *the X-15 program*.

Now, the main part begins: a series of iterative requests to the user to classify sets of URIs. In each iteration *Keval* reevaluates quality measure of each graph pattern, taking into account set of already classified URIs. Quality measure may be any function aggregating the numbers of True Positives, False Positives, True Negatives and False Negatives. For each pattern, True (False) Positives are relevant (resp. irrelevant) URIs that match this pattern, while True (False) Negatives are irrelevant (resp. relevant) URIs that do not match the pattern. The number of URIs matching pattern, but unknown to be relevant or irrelevant, is not taken into account, because we do not want to penalize patterns for suggesting new URIs (after all, that is the point of this algorithm). There were proposed several measures aggregating these four values. Currently we use harmonic F-Measure [8], but we plan to investigate other measures as well, as some authors criticize this measure [11].

In each iteration, *Keval* presents to the user a set of URIs to be classified as relevant or irrelevant. The size of this set is limited by a value specified by the user in the beginning and should be low enough to allow her to reliably classify all of the presented URIs. Selected for the presentation may be only URIs not yet classified and matching a subset of patterns selected by the highest value of the quality measure. Currently this subset is a predefined fraction of the whole set of patterns matching at least one unclassified URI, but other strategies will be also researched in the future. When it is possible, an equal number of URIs for each pattern is presented. When it is not possible or there are any places left, the URIs supported by the most patterns are selected. In case of a tie, the choice is done randomly.

For example, in the first iteration patterns matching Apollo Program astronauts have high quality, because the only URIs known to be relevant are matched by them. For this reason, the algorithm presents to the user a list of astronauts, not only from Apollo 11. The user must now decide which URIs are relevant for her. These, which are not selected as relevant, are assumed to be irrelevant. If she decides that none are relevant, the algorithm stops and the user is presented with the set of patterns selected for their best quality (including possibly patterns with all matching URIs classified). Otherwise, the next iteration of the algorithm is executed: the quality of the patterns is computed based on the new sets of URIs, new examples are generated for the user, etc. In our example, the user selects as relevant `dbr:Michael_Collins_(astronaut)` and `dbr:John_Young_(astronaut)`, and the other presented URIs are by default classified as irrelevant. While the relevant set still contains only astronauts, the irrelevant set now also contains them, and the quality measure of astronaut-related patterns is now decreased.

In the second iteration Apollo 11-related patterns are found to not be significantly worse than astronaut-related ones and the user gets a list containing both astronauts and things related to Apollo 11. She selects `dbr:Tranquility_Base` as relevant. Apollo 11-related patterns now gets even higher quality than others.

In the third iteration, the most of the presented URIs match pattern `?uri dct:subject dbc:Apollo_11`. In the next iteration, the algorithm selects a pattern with lower quality, as the aforementioned patterns are matched only by already classified URIs. For this reason, in the fourth iteration, the user gets a list of URIs that are not interesting for her, so she selects none of them, and algorithm stops. A pattern `?uri dct:subject dbc:Apollo_11` is not the only result - *Keval* has found also patterns related to a book *The first on the Moon*, a movie *Moonwalk One* and *Cullum Geographical Medal*. All of the patterns are consistent with the preferences of the user, but provide new knowledge.

4   Conclusions and Future Work

In this paper, we presented an algorithm allowing the user to mine basic graph patterns describing a given set of URIs. The algorithm operates by interactively asking the user to evaluate intermediate results to narrow down the set of patterns. The algorithm has a few components, that can be fine tuned: pattern quality measure, a strategy for merging triple patterns into Basic Graph Patterns and a strategy of filtering the best patterns. For example, our preliminary results show that using *recall* leads to presenting a lot of URIs irrelevant to the user (for this reason we currently use F-Measure instead). In the future, we would like to analyze different measures and research how used measure influences the interaction with the algorithm: number of presented URIs, number of iterations and quality of the obtained patterns.

Acknowledgement. The work was supported by Polish National Science Center, grant DEC-2013/11/N/ST6/03065.

References

1. Campinas, S.: Live SPARQL auto-completion. In: Horridge, M., Rospocher, M., van Ossenbruggen, J. (eds.) Proceedings of the ISWC 2014 P&D Track. CEUR Workshop Proceedings, vol. 1272, pp. 477–480 (2014)
2. Ferré, S.: SPARKLIS: a SPARQL endpoint explorer forexpressive question answering. In: Horridge, M., Rospocher, M., van Ossenbruggen, J. (eds.) Proceedings of the ISWC 2014 P&D Track. CEUR Workshop Proceedings, vol. 1272, pp. 45–48 (2014)
3. Lawrynowicz, A.: Grouping results of queries to ontologicalknowledge bases by conceptual clustering. In: Nguyen, N.T., Kowalczyk, R., Chen, S. (eds.) Computational Collective Intelligence. LNCS, vol. 5796, pp. 504–515. Springer, Heidelberg (2009)
4. Lawrynowicz, A., Potoniec, J.: Pattern based feature construction in semantic data mining. Int. J. Semant. Web Inf. Syst. **10**(1), 27–65 (2014)
5. Lehmann, J., Isele, R., et al.: Dbpedia - a large-scale, multilingual knowledge base extracted from wikipedia. Semant. Web **6**(2), 167–195 (2015)
6. Leon, A.D., Wisniewki, F., Villazón-Terrazas, B., Corcho, O.: Map4rdf-faceted browser for geospatial datasets. In: USING OPEN DATA: Policy Modeling, Citizen Empowerment, Data Journalism (2012)
7. Marchionini, G.: Exploratory search: from finding to understanding. Commun. ACM **49**(4), 41–46 (2006). http://doi.acm.org/10.1145/1121949.1121979
8. Nguyen, G.H., Bouzerdoum, A., Phung, S.L.: Learning pattern classification tasks with imbalanced data sets. INTECH Open Access Publisher (2009)
9. Oren, E., Delbru, R., Decker, S.: Extending faceted navigation for RDF Data. In: Cruz, I., Decker, S., Allemang, D., Preist, C., Schwabe, D., Mika, P., Uschold, M., Aroyo, L.M. (eds.) ISWC 2006. LNCS, vol. 4273, pp. 559–572. Springer, Heidelberg (2006). doi:10.1007/11926078_40
10. Paulheim, H., Fürnkranz, J.: Unsupervised generation of data mining features from linked open data. In: Burdescu, D.D., Akerkar, R., Badica, C. (eds.) 2nd International Conference on Web Intelligence, Mining and Semantics, pp. 31: 1–31: 12. ACM (2012)
11. Powers, D.M.W.: What the f-measure doesn't measure: features, flaws, fallacies and fixes. CoRR abs/1503.06410 (2015). http://arxiv.org/abs/1503.06410
12. Prud'hommeaux, E., Seaborne, A.: SPARQL query language for RDF: W3C recommendation, W3C, January 2008. http://www.w3.org/TR/2008/REC-rdf-sparql-query-20080115/
13. e Zainab, S.S., Saleem, M., et al.: Fedviz: a visual interface for SPARQL queries formulation and execution. In: Ivanova, V., Lambrix, P., et al. (eds.) Proceedings of the International Workshop on Visualizations and User Interfaces for Ontologies and Linked Data. CEUR Workshop Proceedings, vol. 1456, p. 49 (2015)

Swift Linked Data Miner Extension for WebProtégé

Tomasz Sosnowski, Jedrzej Potoniec[(✉)], and Agnieszka Ławrynowicz

Faculty of Computing, Poznan University of Technology,
ul. Piotrowo 3, 60-965 Poznan, Poland
tomasz.sosnowski@student.put.poznan.pl,
{jpotoniec,alawrynowicz}@cs.put.poznan.pl

Abstract. *Swift Linked Data Miner* (SLDM) is a data mining algorithm capable to infer new knowledge and thus extend an ontology by mining a Linked Data dataset. We present an extension to *WebProtégé* providing SLDM capabilities in a web browser. The extension is open source and readily available to use.

Keywords: *WebProtégé* · *Swift Linked Data Miner* · Linked Data · Data mining · SPARQL

1 Introduction

Swift Linked Data Miner (SLDM) is an algorithm designed for discovering frequent patterns in a Linked Data dataset, i.e. an RDF graph [4]. It is able to infer potentially new and unknown knowledge from the patterns that have been found, thus suggesting to the user possible new axioms that may be added to an ontology. The idea of SLDM is based on inductive reasoning combined with recursive exploration of the graph, so the resulting axioms are not guaranteed to be true in every case, but may provide a useful generalization for the given graph. The results of the mining may be taken into account for extending the existing ontology in order to describe things in a more ordered way. SLDM accesses the graph via a SPARQL endpoint chosen by the user. It starts with a named class, retrieves from the endpoint its members and recursively explores them in order to discover patterns, that may be expressed as OWL 2 EL superclass expressions.

In order to make SLDM available for more researchers, we integrated it with *WebProtégé*, a popular online tool for ontology management [5]. To avoid problems with server overload, SLDM is run as client-side code, which is executed by the Web browser of the user, so no installation of additional software is necessary.

For the pattern detection, SLDM uses a set of rules and thus is capable of finding the following types of patterns:

functional pattern every subject has no more than 1 instance of given predicate, e.g. a date of birth;

datatype property pattern a predicate leads to an RDF literal of a particular datatype;

P. Ciancarini et al. (Eds.): EKAW 2016 Satellite Events, LNAI 10180, pp. 184–187, 2017.
DOI: 10.1007/978-3-319-58694-6_28

range pattern a predicate leads to an RDF literal with a value from a particular range, e.g. an integer value denoting a month is between 1 and 12;

concrete pattern a given predicate always leads to the same object.

2 Extension Description

Swift Linked Data Miner is implemented in *Java* using *Apache Jena* to query SPARQL endpoint [3] and OWL API to manage the results [1]. *WebProtégé* front end uses *Google Web Toolkit* (GWT) framework [5], which compiles *Java* code into *JavaScript* that runs in a web browser. Unfortunately, GWT is able to emulate only a subset of the *Java Runtime Environment*, which does not support *Apache Jena* at all and only a too limited subset of OWL API[1]. Due to these restrictions, a significant part of SLDM has been reimplemented with use of own classes for storing RDF terms and triples in order to meet GWT requirements and replace Jena framework.

The SLDM extension adds to *WebProtégé* a new portlet, available in the *Add content* menu. Figure 1 shows a user interface of the extension, with a text field

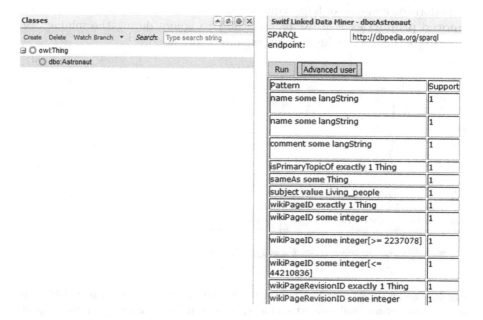

Fig. 1. After the user selected class *Astronaut* (http://dbpedia.org/ontology/ Astronaut) in the class tree (1), the portlet title has changed (2). The *Run* button (3) is now active and allows the user to launch the algorithm using the specified SPARQL endpoint (4). The result is a list of discovered patterns (in form of class expressions) and their supports (5). The *Advanced user* button opens an extended interface shown in Fig. 2.

---

[1] https://github.com/matthewhorridge/owlapi-gwt.

for the URL of a SPARQL endpoint to use. To get the axioms using the basic interface the user should select a class in the class tree portlet and then use the *Run* button.

An extended interface for an advanced user is also provided, as shown in Fig. 2. There, the user may specify the following parameters:

SPARQL endpoint specifies an URL used to access the RDF graph
initial query is a query that provides a set of IRIs used by SLDM as a starting point for a recursive mining of the RDF graph
minimal support is a minimal ratio of triples supporting given pattern to all triples in the investigated set containing a particular predicate
minimal functional support analog to *minimal support*, but used only for mining functional patterns
maximal level limits the maximal overall depth of recursion, thus implying maximal complexity of mined patterns
maximal open level limits the maximal depth of recursion without mining any new pattern
maximal number of endpoint responses limits the number of query results at each recursion level by using SPARQL LIMIT clause and may improve performance in case of large, complex graphs
ignored properties is a list of predicates to ignore during the mining, for example these conveying only provenance information.

Fig. 2. The advanced user interface of the SLDM portlet, showing a text field with default value of minimal support and a separate field for minimal functional support (1). Also shown are the values of maximal level (2), maximal open level (3), maximal number of endpoint responses (by default – no limit) (4) and a text area for ignored properties (5).

3   Proposed Demo

We will present *WebProtégé* with SLDM extension querying over a DBpedia dataset [2], accessed by own SPARQL endpoint. The spectators will be able to observe how various settings of the parameters affect the performance. A short video showing *WebProtégé* SLDM in action is available at https://www.youtube.com/watch?v=2xJPbZVcabQ. During the demo a similar setup will be presented.

4   Conclusion

In this paper we presented an extension to *WebProtégé* providing an interface to *Swift Linked Data Miner*, an algorithm capable of inferring new knowledge from Linked Data datasets available on the Internet. To avoid overloading a server hosting *WebProtégé*, *Swift Linked Data Miner* has been partially rewritten and recompiled with GWT into *WebProtégé* client-side code. Thanks to the basic user interface SLDM is a data mining solution comprehensible even for inexperienced users. The browser-based technology allows for easy use in a friendly environment.

Acknowledgement. This work was partially supported by the PARENT-BRIDGE program of Foundation for Polish Science, co-financed from European Union, Regional Development Fund (Grant No. POMOST/2013-7/8) and by Polish National Science Center, grant DEC-2013/11/N/ST6/03065.

References

1. Horridge, M., Bechhofer, S.: The OWL API: a Java API for OWL ontologies. Semant. Web **2**(1), 11–21 (2011)
2. Lehmann, J., Isele, R., Jakob, M., Jentzsch, A., Kontokostas, D., Mendes, P.N., Hellmann, S., Morsey, M., van Kleef, P., Auer, S., Bizer, C.: DBpedia - a large-scale, multilingual knowledge base extracted from Wikipedia. Semant. Web **6**(2), 167–195 (2015). http://dx.doi.org/10.3233/SW-140134
3. McBride, B.: Jena: a semantic web toolkit. IEEE Internet Comput. **6**(6), 55–59 (2002)
4. Potoniec, J., Jakubowski, P., Ławrynowicz, A.: Swift Linked Data Miner: anytime algorithm for mining OWL 2 EL class expressions directly from on-line linked data. J. Web Semant. https://goo.gl/HFghXp
5. Tudorache, T., Nyulas, C., Noy, N.F., Musen, M.A.: Webprotégé: a collaborative ontology editor and knowledge acquisition tool for the web. Semant. Web **4**(1), 89 (2013)

Exposing RDF Archives Using Triple Pattern Fragments

Ruben Taelman[(✉)], Ruben Verborgh, and Erik Mannens

imec - Ghent University - IDLab, Sint-Pietersnieuwstraat 41, 9000 Ghent, Belgium
{ruben.taelman,ruben.verborgh,erik.mannens}@ugent.be

Abstract. Linked Datasets typically change over time, and knowledge of this historical information can be useful. This makes the storage and querying of Dynamic Linked Open Data an important area of research. With the current versioning solutions, publishing Dynamic Linked Open Data at Web-Scale is possible, but too expensive. We investigate the possibility of using the low-cost Triple Pattern Fragments (TPF) interface to publish versioned Linked Open Data. In this paper, we discuss requirements for supporting versioning in the TPF framework, on the level of the interface, storage and client, and investigate which trade-offs exist. These requirements lay the foundations for further research in the area of low-cost, Web-Scale dynamic Linked Open Data publication and querying.

Keywords: Linked Data · Versioning · Triple Pattern Fragments · Linked Data Fragments · SPARQL

1 Introduction

The ability to perform both simple and complex queries over Linked Datasets is of utmost importance for gaining insights into data. Data analysis can be time-agnostic, but historical and real-time analysis requires probing data *at* certain points in time, or *over* time periods. For example, retrieving biomedical patient information at a certain point in time, or analyzing the evolution of a disease over time.

RDF [1] and SPARQL [5] allow us to represent and query Linked Data. Most Linked Datasets are dynamic on dataset, schema and/or instance level [6]. These refer to additions, changes or deletions of respectively complete datasets, ontologies and separate facts.

A survey on archiving Linked Open Data [3] motivates the relevance and importance of this domain. The authors point out that the current approaches have scalability drawbacks because they are not sufficiently designed or applied to the required Web-Scale. They consider an efficient solution as having a scalable model for archiving, efficient compression and indexing methods which should support a sufficiently expressive temporal query language. In order to expose such dynamic data at a Web-Scale, low-cost publication techniques like Triple

© Springer International Publishing AG 2017
P. Ciancarini et al. (Eds.): EKAW 2016 Satellite Events, LNAI 10180, pp. 188–192, 2017.
DOI: 10.1007/978-3-319-58694-6_29

Pattern Fragments (TPF) [9] could be used. Federated archive quering, which can not be done efficiently with current solutions, would become possible since the TPF framework supports this natively.

In this paper, we discuss the addition of versioning capabilities for instance-level changes to the lightweight TPF interface to reduce the publication and query cost of versioned Linked Open Data. This includes the requirements for such a solution and a research plan for future work.

2   Related Work

A survey about archiving Linked Open Data [3] categorizes instance-level archive solutions into three non-orthogonal storage strategies.

1. The *Independent copies* (IC) approach creates a separate instantiations of datasets for each change or set of changes.
2. The *Change-based* (CB) approach instead only stores changes between versions.
3. The *Timestamp-based* (TB) approach stores the temporal validity of facts.

Five foundational non-orthogonal query atoms for querying archiving systems were introduced [4].

1. *Version materialization* (VM) retrieves data using queries targeted at a single version.
2. *Delta materialization* (DM) retrieves query result differences between two versions.
3. *Version query* (VQ) annotates query results with the versions in which they are valid.
4. *Cross-version join* (CV) joins the results of two queries between versions.
5. *Change materialization* (CM) returns a list of versions in which a given query produces consecutively different results.

Triple Pattern Fragments (TPF) [9] is framework based on a REST interface for querying Linked Data. It was introduced as an alternative to expensive SPARQL endpoints. The TPF interface only allows single triple pattern queries, where linked pages contain the resulting triples. Full SPARQL queries are evaluated by clients, which split them into triple pattern queries, sending those to the server, and joining the results together locally.

3   Requirements

In order to add versioning support to TPF, we place requirements on the server interface, the server's internal storage model and the client that interacts with the interface.

3.1   Interface

The server interface influences the server load when publishing Linked Data archives. More complex interfaces lead to a higher number of possible requests and a lower level of cacheability. The TPF interface has parameters for triple pattern *subject, predicate* and *object*. In order to support the five foundational archive query atoms, this interface needs to be extended. It is however not required to support every one of these query atoms separately, since clients could derive some from others, at the cost of additional computation. For example, DM queries can be resolved using a VM interface by performing two VM queries and calculating the difference in results.

When extending a TPF interface for VM queries, we can expose a list of all available versions as response metadata or through a separate control. These versions could either directly link to interfaces at different versions [8] or those versions can be used as input to an extended interface with a *version* parameter. For example, a request for s1 p1 ? to version 5 using the second method could be http://example.org/?s=s1&p=p1&version=5. DM queries similarly require a *version_initial* and *version_final* parameter to define the version range over which differences should be retrieved. A version query does not require direct changes to the interface, since all results are directly annotated with the versions in which they are valid [7]. CV queries need two times the set of parameters for VM queries, where the two sets define the queries to be joined server-side. CM requires a separate interface with parameters for a single triple pattern, which returns a list of versions for which the given triple pattern produces consecutively different results.

The choice of supported query atoms is a trade-off in server-client load, which can be illustrated using the Linked Data Fragments axis [9]. For a low client effort, it would be ideal for the server to support all five query atoms, leading to a higher server effort. If a low server effort is desired, only one of the first four query atoms should be supported, since the client can calculate any other query based on that.

3.2   Storage

Behind the server interface, there has to be a storage solution that is able to handle chosen query atoms. IC, CB and TB are three approaches for building such a storage solution.

Experimental results show that there exists a trade-off between the storage strategy, query efficiency and storage size [4]. If for example only VM queries are required, the IC policy might be selected if storage size is not an issue. Otherwise, the TB or a hybrid IC-CB approach might be more appropriate at the cost of an increase in server complexity.

Inspired by the classification of complexity for storage policies [3], Table 1 shows qualitative levels of complexity for each query atom. This shows that the supported query atoms at the TPF interface influence the complexity of the storage strategy.

Table 1. Complexity (low ○, medium ◐, high ●) for evaluating **query atoms (rows)** on storage policies (columns).

	IC	CB	TB
VM	○	●	◐
DM	●	○	◐
VQ	◐	◐	○
CV	●	○	◐
CM	●	◐	◐

Table 2. The number of queries required for evaluating **query atoms (rows)** using other query atoms (columns).

	VM	DM	VQ	CV	CM
VM	1	n	1	1	
DM	2	1	1	2	
VQ	n	n	1	n	
CV	2	$2n$	1	1	
CM	n	n	1	n	1

Table 3. Complexity (low ○, medium ◐, high ●) for evaluating **query atoms (rows)** using other query atoms (columns).

	VM	DM	VQ	CV	CM
VM	○	◐	◐	○	
DM	◐	○	◐	◐	
VQ	●	●	○	●	
CV	◐	●	◐	○	
CM	●	●	◐	●	○

3.3 Client

A TPF client should be able to understand the extended TPF server interface. For query atoms that are not supported on the server interface, the client should be able to simulate these using one or more supported query atoms. This simulation has a cost in terms of request count and computation, as shown in Tables 2 and 3. This shows that the choice of supported query atoms on the server will have an impact on the client query efficiency.

Users must be able to interact with one or more versions. This can be done by adding temporal query capabilities to the TPF client using archiving [6] or stream query languages [2]. Or by automatically selecting appropriate versions to query against.

4   Conclusions

To conclude, we foresee three tasks for supporting versioning in the TPF framework.

Task 1: Extension of the TPF interface for at least VM, DM, VQ or CV.

Task 2: A storage solution must be chosen depending on the required storage policies and query atoms.

Task 3: The TPF client must be able to consume the TPF interface extension for the desired and supported query atoms. Users can direct this behaviour either implicitly or through an expressive temporal query language.

These tasks entail a trade-off between server and client load, but also between server storage and server complexity.

Two TPF approaches already exists that meet to these requirements. The TPF Memento extension [8] supports VM queries, uses the IC storage policy and allows clients to select a target date for query evaluation. The TPF Query Streamer engine [7] supports VQ queries and stores data using the TB policy. It evaluates queries for a single moment in time.

No single solution will perform better than all other solutions in all cases. Therefore, we need to explore in which cases, which query atoms should be supported by the interface. And which storage policies and solutions are appropriate for the selected query atoms. It is also important to investigate how efficient query can be simulated if only a subset is supported by the interface. A server and client cost model is needed to quantify the trade-off between server and client load, and between server load and storage. This cost model should have parameters for storage policies and the different supported client and server query atoms.

References

1. Cyganiak, R., Wood, D., Lanthaler, M.: RDF 1.1: concepts and abstract syntax. Recommendation, W3C, February 2014. http://www.w3.org/TR/2014/REC-rdf11-concepts-20140225/
2. Dell'Aglio, D., Della Valle, E., Calbimonte, J.P., Corcho, O.: RSP-QL semantics: a unifying query model to explain heterogeneity of RDF stream processing systems. Int. J. Semant. Web Inf. Syst. (IJSWIS) 10(4), 17–44 (2014)
3. Fernández, J.D., Polleres, A., Umbrich, J.: Towards efficient archiving of dynamic linked open data. In: Proceedings of DIACHRON, pp. 34–49 (2015)
4. Fernández, J.D., Umbrich, J., Polleres, A., Knuth, M.: Evaluating query and storage strategies for RDF archives
5. Harris, S., Seaborne, A., Prud'hommeaux, E.: SPARQL 1.1 query language. Recommendation, W3C, March 2013. http://www.w3.org/TR/2013/REC-sparql11-query-20130321/
6. Meimaris, M., Papastefanatos, G., Viglas, S., Stavrakas, Y., Pateritsas, C., Anagnostopoulos, I.: A query language for multi-version data web archives (2015)
7. Taelman, R., Verborgh, R., Colpaert, P., Mannens, E., Van de Walle, R.: Continuously updating query results over real-time Linked Data. In: Proceedings of the 2nd Workshop on Managing the Evolution and Preservation of the Data Web, May 2016
8. Verborgh, R.: Querying history with Linked Data (2016). http://ruben.verborgh.org/blog/2016/06/22/querying-history-with-linked-data/
9. Verborgh, R., Vander Sande, M., Hartig, O., Van Herwegen, J., De Vocht, L., De Meester, B., Haesendonck, G., Colpaert, P.: Triple Pattern Fragments: a low-cost knowledge graph interface for the Web. J. Web Semant. 37–38, 184–206 (2016)

DKA-robo: Dynamically Updating Time-Invalid Knowledge Bases Using Robots

Ilaria Tiddi, Emanuele Bastianelli$^{(\boxtimes)}$, Enrico Daga, and Mathieu d'Aquin

Knowledge Media Institute, The Open University, Milton Keynes, UK
{ilaria.tiddi,emanuele.bastianelli,
enrico.daga,mathieu.daquin}@open.ac.uk

Abstract. In this paper we present the DKA-robo framework, where a mobile robot is used to update the statements of a knowledge base that have lost validity in time. Managing the dynamic information of knowledge bases constitutes a key issue in many real-world scenarios, because constantly reevaluating data requires efforts in terms of knowledge acquisition and representation. Our solution to such a problem is to use RDF and SPARQL to represent and manage the time validity of information, combined with a robot acting as a mobile sensor which updates the outdated statements in the knowledge base, therefore always guaranteeing time-valid results against user queries. This demo shows the implementation of our approach in the working environment of our research lab, where a robot is used to sense temperature, humidity, wifi-signal and number of people on demand, updating the lab knowledge base with time-valid information.

1 Introduction

Managing dynamic data in knowledge bases, i.e. statements that are only valid for a certain period of time, is a well-known problem for knowledge acquisition and representation, because data need to be constantly re-evaluated to allow reasoning. Our current work focuses on representing the validity in time of statements in a knowledge base, and on using an autonomous robot as a mobile sensor that updates the outdated statements, therefore constantly guaranteeing the "freshness" of the requested information.

Let us imagine a knowledge base representing the working environment of the Knowledge Media Institute (KMi) research department, where information about locations is static (e.g. the coordinates of a room) but information such as temperature, humidity, wi-fi signal or number of people in a room varies more often and needs to be re-evaluated. In scenarios where scale, lack of infrastructure or accessibility are a challenge, existing solutions to this problem, such as providing time-stamped versions of the knowledge base [3,4,7] or using sensors to constantly stream the information [1,2,5], might not be achievable at reasonable costs, and might provide information that is likely not to be required.

The alternative solution of moving an autonomous robot upon request to re-collect the expired information (i.e. that has lost time validity) and update

© Springer International Publishing AG 2017
P. Ciancarini et al. (Eds.): EKAW 2016 Satellite Events, LNAI 10180, pp. 193–197, 2017.
DOI: 10.1007/978-3-319-58694-6_30

the knowledge base has the advantage of guaranteeing that, when queried, the knowledge base will always return time-valid information. In this context, there is a number of challenges to be faced at knowledge representation and management level, namely: how to represent time validity in the knowledge base; how to establish that statements have expired; and how to instruct a robot to perform a set of actions by favouring the time validity of the information that is collected.

Our solution is to use RDF and SPARQL as a framework for knowledge representation, combined with an autonomous robot which updates the time-invalid information of the knowledge base on demand. More specifically, our tool uses statements in the knowledge base with a time-stamp representing their time validity (i.e. their expiry date). When the knowledge base is queried through SPARQL, we decide how long a piece of information necessary to answer this query will be considered valid. On this basis, the tool creates a plan for the robot to collect such information, in a way that guarantees the time validity of the information returned.

This demo, that we call DKA-robo (Dynamic Knowledge Acquisition with a Robot), presents the implementation of our approach based on the simulated working environment of our research lab. Users will be shown how the robot can be instructed to move and sense information, and how the evaluated plan is executed to preserve the time validity of our knowledge base.

2 Process Overview

DKA-robo is implemented as a process in which a user submits a query to a knowledge base and receives a set of results that are valid in time, provided that there exists a plan that enables the robot to collect the required information in the limited time before it becomes invalid. The process, described more in detail in [6], is articulated as reported below.

1. **Query.** The user expresses the query to the knowledge base. The knowledge base is represented as a set of RDF quads $q = (t, g)$, where t is the ⟨subject, predicate, object⟩ triple and g is the named graph to which t belongs, representing the time at which t will expire.
2. **Invalid information collection.** The query is first executed onto the current knowledge base, and the execution process is monitored to retrieve the graphs from which triples are obtained. From these, triples which have expired and which have not are identified.
3. **Planning.** The planner receives the time-invalid quads and asks the robot for its location. Based on these, it calculates the plan to send to the robot, i.e. the right sequence of actions to perform so that none of the statements in the answer set remains time-invalid. The plan is evaluated using a best-first strategy designed to minimise the time to invalidity of the considered quads[1].

---

[1] Since we focus on the representation of time validity in the knowledge base, we employ a naïve implementation of the planner without claiming for its efficiency.

4. **Knowledge base update.** The robot receives the plan and performs it. When a new piece of information is collected, it is sent to the knowledge base to be updated. The time to invalidity of the new information is evaluated using a set of time validity rules expressed as a triple pattern p associated to a duration d. The rule for which the triple t matches the pattern p that has the shortest duration is selected, and a new quad $q = (t, \text{current\_time}+d)$ is written in the knowledge base.

5. **Query results.** Once all the actions have been executed, the user is shown the answer to its query. If no answer is received, this means that there is no plan that can be executed such that all the statements in the answer are time-valid.

3   DKA-robo Demonstration

During the demonstration, we will present scenarios in a simulated environment from a real-world example, where a robot moves in KMi on demand, updating the outdated information of the knowledge base. The audience will be able to interact with DKA-robo through the user interface shown in Fig. 1 and described below.

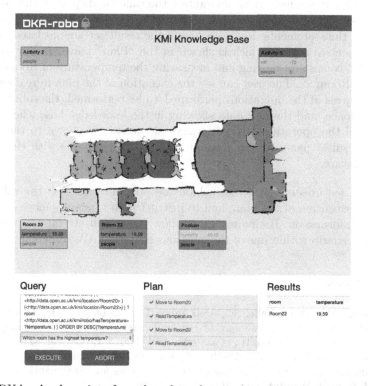

Fig. 1. DKA-robo demo interface: the coloured areas represent rooms and activities in the KMi.

In the first part, shown at the top of Fig. 1, the user is shown the KMi knowledge base and its status as a map. The dynamic information of each room (temperature, humidity, wi-fi strength in dB and number of people) fades out the more the information is approaching its expiry date. For instance, Fig. 1 shows that information about the Podium and Activity 5 are about to expire. These will reappear when the robot will send the new sensed information. The user can also see the location of the robot on the map while executing the plan (in Room 22 in our example).

The second part, shown at the bottom in Fig. 1, is dedicated to the user's interaction with both the knowledge base and the robot. In the "Query" panel, users can freely query the knowledge base, either by choosing one of the predefined queries expressed in natural language, or by inserting their own SPARQL query. In Fig. 1, a user asked for the room with the highest temperature. Given a query, three scenarios are possible:

1. no outdated statement is used to compute the result set. No plan is sent to the robot and the result set is shown directly in the "Results" panel.
2. some of the statements used to compute the result set are outdated, but the robot receives no plan. This means that there is no possible plan to be found by the planner, which guarantees the time validity of all the necessary statements. The user is therefore alerted that the query has to be simplified.
3. some of the statements used to compute the result set are outdated, and the robot receives a plan, which is shown in the "Plan" panel. In our example, the plan consists in moving and measuring the temperature of Room 20 and then of Room 22. The user can see the execution of the plan in real-time, i.e. the progress of the operations performed/to be performed, the robot moving in the space, and the new data showing in the knowledge base when sensed. Once all the operations are performed, the answer is shown to the user in the "Results" panel. In our example, Room 20 is the one with the highest temperature.

Additional features of the demo include the possibility to stop the robot and abort the execution of the plan, and to instruct it to randomly move and sense outdated information. If circumstances allow it, we will attempt to reproduce the KMi scenario within one of the locations of the EKAW conference.

References

1. Balduini, M., Della Valle, E., Dell'Aglio, D., Tsytsarau, M., Palpanas, T., Confalonieri, C.: Social listening of city scale events using the streaming linked data framework. In: Alani, H., et al. (eds.) ISWC 2013. LNCS, vol. 8219, pp. 1–16. Springer, Heidelberg (2013). doi:10.1007/978-3-642-41338-4_1
2. Calbimonte, J.-P., et al.: Enabling query technologies for the semantic sensor web. Int. J. Semant. Web Inf. Syst. 8(1), 43–63 (2012)
3. Fernández, J.D., et al.: The DBpedia wayback machine. In: The 11th International Conference on Semantic Systems. ACM (2015)

4. Halpin, H., Cheney, J.: Dynamic provenance for SPARQL updates using named graphs. In: 23rd International Conference on World Wide Web (2014)
5. Le-Phuoc, D., Dao-Tran, M., Pham, M.-D., Boncz, P., Eiter, T., Fink, M.: Linked stream data processing engines: facts and figures. In: Cudré-Mauroux, P., et al. (eds.) ISWC 2012. LNCS, vol. 7650, pp. 300–312. Springer, Heidelberg (2012). doi:10.1007/978-3-642-35173-0_20
6. Tiddi, I., et al.: Update of time-invalid information in knowledge bases through mobile agents. In: Integrating Multiple Knowledge Representation and Reasoning Techniques in Robotics (2016)
7. Van de Sompel, H., et al.: An HTTP-based versioning mechanism for linked data. arXiv preprint (2010). arXiv:1003.3661

Selecting Documents Relevant for Chemistry as a Classification Problem

Zhemin Zhu$^{(\boxtimes)}$, Saber A. Akhondi, Umesh Nandal, Marius Doornenbal,
and Michelle Gregory

Elsevier, Radarweg 29, 1043 NX Amsterdam, The Netherlands
{z.zhu1,s.akhondi,u.nandal,m.Doornenbal,m.gregory}@elsevier.com

Abstract. We present a first version of a system for selecting chemical publications for inclusion in a chemistry information database. This
database, Reaxys (https://www.elsevier.com/solutions/reaxys), is a por-
tal for the retrieval of structured chemistry information from published
journals and patents. There are three challenges in this task: (i) Training
and input data are highly imbalanced; (ii) High recall ($\geq$95%) is desired;
and (iii) Data offered for selection is numerically massive but at the same
time, incomplete. Our system successfully handles the imbalance with
the undersampling technique and achieves relatively high recall using
chemical named entities as features. Experiments on a real-world data
set consisting of 15,822 documents show that the features of chemical
named entities boost recall by 8% over the usual n-gram features being
widely used in general document classification applications. For fostering
research on this challenging topic, a part of the data set compiled in this
paper can be requested.

Keywords: Natural language processing · Document classification ·
Machine learning · Cheminfomatics

1 Introduction and Problem Statement

Publications including articles and patents in chemistry contains valuable information of newly developed *compounds* and their *properties* and *relations* with
other chemicals [5]. Publications concerning chemistry has grown exponentially
from 450 k publications in 1995 to over 24 million, with over 1.1 million articles
published in 2014 alone. Thus automatically extracting structured information
from chemistry literature has become a critical step for novelty checking, patent
validation, finding new starting points for chemical research, especially in drug
discovery research in pharmaceutical industry. This provides a big arena and
also poses substantial challenges for NLP and ML techniques [4,8].

At present, there are a few chemical databases that contain chemistry information which were manually excerpted from literature. For example, Elsevier's
Reaxys incorporates manually excerpted compounds, their properties and reac-
tions from selected patents and journals. But with the fast growing number of

© Springer International Publishing AG 2017
P. Ciancarini et al. (Eds.): EKAW 2016 Satellite Events, LNAI 10180, pp. 198–201, 2017.
DOI: 10.1007/978-3-319-58694-6_31

publications, manual excerptions are no longer affordable. Manual excerption is also tedious and an error-prone process. In this study, as an initial attempt to automate the manual excerption process, we present a classification method to establish the relevancy of chemistry-related articles for manual excerption.

The problem to be investigate can be formalized as a binary classification task. We classify articles from chemistry journals into two categories: *relevant* and *irrelevant*. Relevant articles will go further for manual excerption, and irrelevant articles will be abandoned. Also in this task we do not want to lose any valuable information, recall takes priority over precision.

2 Dataset

The dataset (Table 1) for experiments was compiled from our manual excerption database, and strictly validated by domain experts.

Table 1. Data sets

Chemistry area	#Rel.	#Irrel.	#Total
Physical Chemistry	876	1,280	2,156
Medicinal Chemistry	195	5,985	6,180
Biochemistry	865	2,107	2,972
Unspecified	2,253	2,261	4,514
Total	4,189	11,633	15,822

The data are imbalanced. In total, there are many more irrelevant articles $(11,633)$ than relevant ones $(4,189)$. While this implies that there is a big potential for saving excerption effort by filtering, it also brings a technical challenge to the classifier to handle imbalanced data.

3 Approach

Liblinear [2] was employed as the binary classifier. Liblinear supports linear support vector machines and is scalable to large data sets. Moreover, it can be efficiently trained with large number of features. In the experiments, we set Liblinear to the type 1 solver: L2-loss support vector classification with the L2 regularization. For tokenization and lemmatization we applied ClearTK [6]. Stopwords were annotated with respect to a stopword list. Finally, in order to organize components as a pipeline and share annotations between components, we employed UIMA framework[1].

Features used in our experiments can be grouped into three categories:
NLP features: For general NLP features, we used shallow n-gram features. We extracted 1 to 4 n-gram features with respect to lemmas from titles and abstracts of articles.

---

[1] https://www.uima.apache.org.

Metadata related features: Metadata, such as journal (source) names and citation types, were included in the feature sets.

Chemistry related features: We employed the chemical compound recognizer provided by OntoChem [3] to extract chemical compounds and use the extracted compounds as features. We used the SMILES representation [9] provided together with the extracted compounds as feature labels.

Additionally, due to the high feature dimensionality, we assessed the performance of the classifier by applying feature reduction techniques such as mutual information (MI) [7].

In order to handle the data imbalance problem during training, we applied the over- and under-sampling strategies [1].

4   Results

Table 2 shows the experimental results. We built five classifiers based on different feature sets and evaluated the performance of each classifier using 5-fold cross-validation (Table 2). We used the complete data i.e. 4,189 *relevant* and 11,633 *irrelevant* (Table 1) to train each classifier. The results obtained with and without undersampling are listed in separate rows. We investigated the significance of different feature sets on the performance of the classifier by firstly, using all features (AF) and then by eliminating each feature type separately. For example, in the feature set AF-nGram, we used the all features except the n-gram features.

Table 2. Classifier evaluation

Feature sets	Undersampling	Relevant			Irrelevant		
		P	R	F1	P	R	F1
All features(AF)	No	0.769	0.620	0.686	0.872	0.933	0.901
	Yes	0.595	0.842	0.697	0.933	0.794	0.858
AF-nGram	No	0.702	0.531	0.605	0.845	0.919	0.880
	Yes	0.475	**0.926**	0.628	0.960	0.631	0.762
AF-citationType	No	0.769	0.616	0.684	0.871	0.933	0.901
	Yes	0.588	0.840	0.692	0.932	0.788	0.854
AF-source	No	0.769	0.615	0.684	0.871	0.933	0.901
	Yes	0.591	0.840	0.694	0.932	0.790	0.855
AF-chemicalCompounds	No	0.769	0.620	0.686	0.872	0.933	0.901
	Yes	0.595	0.845	0.698	0.934	0.793	0.858
Feature reduction on best feature set (AF-nGrams)							
Mutual information	Yes	0.476	0.919	0.627	0.956	0.635	0.763

From Table 2, we can see that the undersampling technique significantly boosts recall on the relevant class. As we explained before, the recall on relevant class is the priority in our task, because we do not want to miss relevant articles. The undersampling technique also slightly improves F1 scores in all feature settings. The best recall (92.6%) on the relevant class is obtained when we leave out n-gram features, and train the classifier with the remaining features: citation type, source and chemical compounds features. Comparing with the best recall obtained while including n-gram features, which is 84.5%, the improvement is significant (8%).

Finally, we applied the mutual information feature selection to the best setting, i.e., using feature sets AF-nGrams with undersampling. However, mutual information feature selection does not improve the recall on the relevant class. The difference between applying and not applying mutual information feature selection is small.

5 Conclusion

We built an initial classification system for classifying chemistry articles. By applying the undersampling technique together with features of chemical named entities, our system achieves 92.6 recall on relevant articles.

References

1. Borrajo, L., Romero, R., Iglesias, E.L., Marey, C.R.: Improving imbalanced scientific text classification using sampling strategies and dictionaries. J. Integr. Bioinform. **8**(3), 176 (2011)
2. Fan, R.E., Chang, K.W., Hsieh, C.J., Wang, X.R., Lin, C.J.: LIBLINEAR: a library for large linear classification. J. Mach. Learn. Res. **9**, 1871–1874 (2008). http://dl.acm.org/citation.cfm?id=1390681.1442794
3. Irmer, M., Lutz, W., Böhme, T., Püschel, A., Claudia, B., Ulf, L.: OCMiner for patents: extracting chemical information from patent texts (2013). http://www.biocreative.org/media/store/files/2015/BCV2015_paper_57.pdf
4. Jessop, D.M., Adams, S.E., Murray-Rust, P.: Mining chemical information from open patents. J. Cheminform. **3**(1), 40 (2011). http://jcheminf.springeropen.com/articles/10.1186/1758-2946-3-40
5. Muresan, S., Petrov, P., Southan, C., Kjellberg, M.J., Kogej, T., Tyrchan, C., Varkonyi, P., Xie, P.H.: Making every SAR point count: the development of chemistry connect for the large-scale integration of structure and bioactivity data. Drug Disc. Today **16**(23–24), 1019–1030 (2011). http://www.sciencedirect.com/science/article/pii/S1359644611003448
6. Ogren, P.V., Wetzler, P.G., Bethard, S.: Cleartk: a UIMA toolkit for statistical natural language processing. In: Towards Enhanced Interoperability for Large HLT Systems: UIMA for NLP 32 (2008)
7. Peng, H., Long, F., Ding, C.: Feature selection based on mutual information criteria of max-dependency, max-relevance, and min-redundancy. IEEE Trans. Pattern Anal. Mach. Intell. **27**(8), 1226–1238 (2005)
8. Vazquez, M., Krallinger, M., Leitner, F., Valencia, A.: Text mining for drugs and chemical compounds: methods, tools and applications. Mol. Inform. **30**(6–7), 506–519 (2011). http://doi.wiley.com/10.1002/minf.201100005
9. Weininger, D.: SMILES, a chemical language and information system. 1. Introduction to methodology and encoding rules. J. Chem. Inform. Model. **28**(1), 31–36 (1988). http://dx.doi.org/10.1021/ci00057a005

Doctoral Consortium

Addressing Knowledge Integration
with a Frame-Driven Approach

Luigi Asprino[(✉)]

DISI, University of Bologna, Bologna, Italy
luigi.asprino@unibo.it

Abstract. Given a knowledge-based system running virtually forever
able to acquire and automatically store new open-domain knowledge,
one of the challenges is to evolve by continuously integrating new knowledge. This needs to be done while handling conflicts, redundancies and
linking existing knowledge to the incoming one. We refer to this task
with the name *Knowledge integration*. In this paper we define the problem by discussing its challenges, we propose an approach for tackling the
problem, and, we suggest a methodology for the evaluation of results.

1 Introduction

Let us imagine a knowledge-based system running virtually forever and able
to acquire and automatically store new open-domain knowledge. An example
of this kind of systems could be a robot equipped with dialoguing capabilities
that enriches its KB with information acquired from interactions with humans
and environmental sensors. The actions that the robot chooses to perform might
be influenced by the current state of its knowledge base, e.g. an assistive robot
learns the preferences of its user by talking to her/him and consequently it performs the actions s/he prefers. The topic of a human-robot dialogue may spread
from the user's personal memories to news. In order to interact with a user, a
robot's KB must contain the information relevant for the discourse. A possible
solution might be to equip the robot with general-purpose background knowledge such as DBpedia or other commonsense KBs. However, despite constantly
growing, crowd-sourced KBs are not tailored for representing personal information. Furthermore, considering the need of dealing with potentially any domain,
it is unrealistic to pre-define the necessary knowledge. A robot needs a knowledge manager able to evolve the robot's KB with the inputs acquired during its
lifetime.

The main challenge in this scenario is to automatically evolve the KB by
continuously integrating new open-domain knowledge while handling conflicts,
redundancies and linking the existing knowledge with the incoming one. The
term knowledge embodies information at both the intensional (e.g. a TBox)
and the extensional level (e.g. an Abox). Therefore, the knowledge base should
evolve both by including new assertions and by evolving the conceptualization.
The problem can be stated as follows.

© Springer International Publishing AG 2017
P. Ciancarini et al. (Eds.): EKAW 2016 Satellite Events, LNAI 10180, pp. 205–210, 2017.
DOI: 10.1007/978-3-319-58694-6_32

> *Given two (or more) information sources (either structured or unstructured), knowledge integration is the problem of automatically building a knowledge base by extracting the entities from the sources and finding the relations that hold among them.*

The relation has to state: (i) The type of relationship between the entities (e.g. equivalence, subsumption, a domain-specific relation etc.); (ii) When it emerged; (iii) The context where it emerged, e.g. the provenance of the knowledge graphs.

In order to introduce the challenges of the thesis, a preliminary clarification on the terminology is needed. We distinguish a *world entity*, i.e. anything (real, possible or imaginary) in the real world, from an *entity*, i.e. concepts, relations and individuals represented in a knowledge base. Knowledge Integration arises a lot of challenging issues: How different information sources represent the same world entity? Is it possible to automatically find correspondences among same[1] entities coming from different sources? If two entities in the sources represent two different world entities, which is the relationship between them? If two entities contradict each other, is it possible to detect and possibly solve the conflict?

Integrating information from different sources is crucial in today's real-world applications, consequently several communities (e.g. working on databases, web semantics, linguistics etc.) have faced this problem. It has been widely studied from several points of view and many interesting solutions have been proposed. Data Integration, Ontology Matching, Ontology Evolution, Knowledge Fusion, Entity Linking, Co-Reference Resolution, Word Sense Disambiguation are all problems focusing on a particular kind of integration. None of these provides individually a comprehensive solution for the problem. However, it could be addressed by extending and orchestrating techniques developed for specific tasks.

Regardless the way of representing information, that can be delivered either in structured (e.g. relational database) or in unstructured (e.g. plain text) format, human beings express information by reflecting the conceptualization they have in mind. A conceptualization of a stereotyped situation is called *Frame* [4]. Frames are data-structures that can represents facts, like participating to a marriage or being in a certain location. Frames have been mainly being used in linguistics [2] for building lexical databases (e.g. FrameNet); in Natural Language Processing for tasks such as Knowledge Extraction and Semantic Role Labeling; and for Exploration of Encyclopedic Information.

The evidences of frames that emerge from all kinds of knowledge sources suggest to use this cognitive model as a driver for knowledge integration. In my doctoral work I am inquiring the possibility of using this *"remembered framework"* (as defined by Minsky [4]) in the knowledge integration task. In this vision frames constitutes the background knowledge that will serve to ground and merge the incoming knowledge from the various sources.

The rest of the paper is organized as follows. Section 2 discusses the related work. Section 3 presents the approach investigated in this PhD thesis work. Finally, Sect. 4 provides concluding remarks and the research plan.

---

[1] *"Same"* means "that refers to same world entity".

2   Related Work

The closest problem to knowledge integration is *Data Integration (DI)*. From a theoretical perspective [3], data integration can be seen as the problem of combining the data residing at different sources, and providing the user with a unified view (also called *global schema*) of these data. The main task of a data integration system is providing the semantic mapping between the sources and the global schema. Pure data integration solutions cannot be employed for many reasons. In DI, the global schema is modeled on top of the sources instead of automatically emerging from them. Furthermore, global schema and semantic mapping are provided at the scratch line and do not automatically evolve over the time. Finally, despite some semi-automatic approach to schema matching, developing the mapping is still manual work.

Ontology Matching (OM) is the problem of finding correspondences between semantically related entities of ontologies. It provides a solution to the semantic heterogeneity problem in order to allow the semantic interoperability of the data expressed with respect to the matched ontologies. Shvaiko and Euzenat [6] surveyed the state of the art and proposed the future challenges for this field. Most of the automatic matching techniques rely on shallow text and structural similarity of the entities of the ontologies. These techniques provide an excellent starting point and could be extended by introducing a deeper analysis of the entity semantics. However, matching is not enough, to create an integrated KB it is also needed to find other kinds of relations that link the existing knowledge with the new one.

Ontology Evolution (OE) aims at maintaining an ontology up to date with respect to changes in the domain that it models. A recent survey by Zablith *et al.* [7] provides a complete overview of the tasks involved in Ontology Evolution process. Knowledge integration is related to the first three stages of the OE process, i.e.: (i) *detecting the need of evolution*; (ii) *suggesting ontology changes*; (iii) *validating ontology changes*.

Recently, Mongioví *et al.* [5] proposed a novel solution to semantic reconciliation. They reduced the problem to a graph alignment problem. The main contribution of this work is introducing the notion of "global optimization" in the entity matching task. In doing so this approach takes advantage of considering the semantics of the entities within the context of the entire knowledge base (i.e. the entities are no longer compared individually). However, the limit of the approach is that it is highly tailored to knowledge graphs extracted by means of FRED.

3   Proposed Approach

The high-level approach that is investigated in this PhD thesis work is summarized in the following two steps. (i) *Entity expansion*. In order to perform a semantic comparison of two entities there is the need of representing their meaning by resorting to a same background knowledge. A background knowledge (such as WordNet, FrameNet or Framester) provides a set of individuals (i.e. concepts, properties etc.) with a well-defined semantics which can be used to describe the semantics of the incoming entities. In general, the semantics of an entity can be represented by a semantic network that relates individuals of the background knowledge. Each source's entity is "replaced" by a richer graph representing its meaning in terms of the background knowledge. (ii) *Comparison.* A graph-based comparison of the semantic networks representing the meaning of the entities is performed. The aim of this task is to detect the relationship between two semantic networks representing two entities coming from the information sources.

This methodology has been implemented for addressing the problem of semantic heterogeneity among ontologies [1]. Most of the current ontology matching solutions present two main limits: (i) they only partially exploit the natural language descriptions of ontology entities and lexical resources as background knowledge; (ii) they are mostly unable to find correspondences between entities specified through different logical types (e.g. mapping properties to classes). We introduced a novel approach aimed at finding complex correspondences between ontology entities according to the intensional meaning of their models, hence abstracting from their logical types.

In order to deduce the intensional meaning of ontology entities we proposed to analyze the natural language annotations associated with them. In fact, annotations provide humans with insights of the intensional meaning the designer wants to represent with a certain entity. The main idea of this approach is that words used in annotations *evoke* frames that are representative of the intensional meaning of the entity. Evoked frames can be used to describe the semantics of the ontology entity. In other words, the *expansion step* of the above methodology has been implemented by using frames as background knowledge to describe entities' semantics. The frame-based representation of the meaning of ontology entities allows us to treat ontology entities as *multigrade predicates* hence abstracting from their logical type.

The proposed approach is currently being evaluated. We are evaluating the resulting alignments in a both *direct* and *indirect* way. The benchmarks used for assessing ontology matching systems[2] are not able to evaluate the capability of finding complex correspondences among ontology entities with different logical types. In order to accomplish this purpose we are extending the existing benchmarks for ontology matching. On the other hand, we are using the proposed approach in a question answering system for selecting relevant resources answering a given question. The frame occurrences in a question together with the frame-ontology alignment help in formulate the query over the linked data, hence identifying resources that answer the given question.

Evaluation of results. We found only one attempt for developing a gold standard for benchmarking the performance of a semantic reconciliation framework. Mongiovi *et al.* [5] proposed to develop a ground truth for semantic reconciliation by adapting EECB 1.0, the gold standard for Co-Reference Resolution. The focus of Mongiovi *et al.* was on the matching task, but we also need to test

---

[2] OAEI, http://oaei.ontologymatching.org/.

the capability of the framework of instancing new semantic relationship and of solving inconsistencies. Therefore the ground truth can be extended (or a new one can be created) by exploiting crowdsourcing techniques, e.g. by designing a "game with a purpose".

The effectiveness of the methodologies and techniques will be further assessed by employing them in a real scenario. For instance, the results of the thesis could be used for managing the knowledge base of an intelligent agent (e.g. an assistive robot, a question answering system etc.) that decides the actions to perform on the basis of the current state of its KB. In this case, the quality of the of the agent behavior can be considered as representative of the effectiveness of the thesis results.

4   Research Plan and Conclusion

This PhD work is dimensioned for three years. The beginning of the first year has been devoted to an intense literature review of the related research areas and to selecting the promising methodologies that could contribute to the development of a solution. The first year has been also devoted to (i) devising the high-level approach for addressing knowledge integration; (ii) and to implementing a novel solution for ontology matching that uses frames as background knowledge for reducing semantic heterogeneity among entities of different ontologies. This solution tackles knowledge integration at *schema level* but in the next future we plan to extend the approach to face the integration at *extensional level*. We hypothesize that the semantics of an individuals can be described through a frame-based specification thus reducing the semantic heterogeneity among individuals of different knowledge graphs.

In the second year we plan to evaluate the approach presented in Sect. 3 (i) by extending the existing benchmarks for ontology matching; (ii) by employing the frame-based ontology alignment in a question answering system as support for creating queries from the input questions.

This paper presented a summary of the PhD thesis in its early stage. We presented the problem of knowledge integration, its challenges and an approach that will be investigated during the next years. Finally, we suggested some methodologies for assessing the effectiveness of a knowledge integration framework.

Acknowledgements. I wish to express my gratitude to my advisors, Prof. Paolo Ciancarini and Dr. Valentina Presutti, for their encouragement throughout my research for this work. I would also like to thank Professor Aldo Gangemi for introducing me to the wonders of the Frame Semantics and for his precious suggestions.

References

1. Asprino, L., Presutti, V., Gangemi, A.: Matching ontologies using a framedriven approach. In: Proceedings of EKAW - Satellite Events, Bologna, Italy (2016)
2. Fillmore, C.J.: Frame semantics. In: Linguistics in the Morning Calm, pp. 111–137. Hanshin Publishing Co. (1982)

3. Lenzerini, M.: Data integration: a theoretical perspective. In: Proceedings of PODS 2002, Madison, Wisconsin, pp. 233–246. ACM (2002)
4. Minsky, M.: A framework for representing knowledge. Technical report, Cambridge, MA, USA (1974). http://hdl.handle.net/1721.1/6089
5. Mongiovì, M., Recupero, D.R., Gangemi, A., Presutti, V., Consoli, S.: Merging open knowledge extracted from text with MERGILO. Knowl. Based Syst. **108**, 155–167. doi:10.1016/j.knosys.2016.05.014
6. Shvaiko, P., Euzenat, J.: Ontology matching: state of the art and future challenges. IEEE Trans. Knowl. Data Eng. **25**(1), 158–176 (2013)
7. Zablith, F., Antoniou, G., d'Aquin, M., Flouris, G., Kondylakis, H., Motta, E., Plexousakis, D., Sabou, M.: Ontology evolution: a processcentric survey. Knowl. Eng. Rev. **30**(1), 45–75 (2015)

Automatic Maintenance of Semantic Annotations

Silvio Domingos Cardoso[1,2](✉)

[1] LIST, Luxembourg Institute of Science and Technology,
5, avenue des Hauts-Fourneaux, 4362 Esch-sur-Alzette, Luxembourg
silvio.cardoso@list.lu
[2] LRI, Univ. Paris-Sud, CNRS, Université Paris-Saclay, Orsay, France

Abstract. Biomedical Knowledge Organization Systems (KOS) play a key role in enriching information in order to make them machine understandable. This is done through semantic annotation which consists in the association of concept labels taken from KOS with pieces of digital information taken from the source to annotate. However, the dynamic nature of these KOS directly impacts on the annotations, creating a mismatch between the enriched data and the concept labels. This PhD study addresses the evolution of semantic annotations due to the evolution of KOS and aims at proposing an approach to automatize the maintenance of semantic annotations.

1 Problem

The use of Knowledge Organization Systems (KOS) [10] such as classification schemes, thesauri or ontologies in the medical field has shown great value to tackle semantic interoperability issues. In many cases, KOS are used to annotate objects such as the content of electronic health records' (EHR), genes or publications in order to make their semantics explicit for software applications. These annotations facilitate the automatic retrieval and exploitation of relevant information. For instance, case report forms (CRF) of clinical trials can be semantically enriched with annotations taken from KOS to improve the analysis of patient profiles at patient recruitment time. The well-known Gene Ontology (GO) is used to describe molecular functions of genes and proteins, and scientific publications in MEDLINE are semantically annotated with concepts of the Medical Subject Headings (MeSH) facilitating the search for relevant medical information [11].

Besides, the dynamic nature of medical knowledge forces knowledge engineers to continuously revise the content of either KOSs or underlying data. For instance, the removal of a concept in a KOS engenders the removal of the semantics of the associated annotation and therefore making it not understandable for computers. More generally, these changes may directly impact the annotations associated with changed concepts or changed data and new KOS versions can potentially invalidate previous annotations. As a result, many annotations can

© Springer International Publishing AG 2017
P. Ciancarini et al. (Eds.): EKAW 2016 Satellite Events, LNAI 10180, pp. 211–218, 2017.
DOI: 10.1007/978-3-319-58694-6_33

lose their relevance and value thus hindering the intended use and exploitation of annotated data.

Consider, for instance the example on Fig. 1. The restructuring of MeSH terminology will impact on the existing annotations, creating a mismatch between the annotations and the KOS. To this end, software applications such as search engines or data portals will not be able to return precise and complete information if the query specifies the concept D025281.

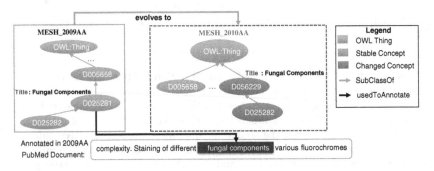

Fig. 1. Annotation evolution case study. A subset of a document is annotated with Fungal components, the *skos:prefLabel* of the concept D025282 in MeSH 2009AA. In the next version the concept D025282 is moved to D056229 in other branch of the ontology. This change has caused a mismatch between the annotation created with the older version and the concept of the new KOS version.

One of the possible solutions to cope with this problem is the re-annotation of documents. However, Funk et al. [7] point that the concept recognition systems vary from ontology to ontology and do not perform equally on natural language texts. Furthermore, the necessity to validate these produced annotations is a laborious and time consuming task for domain specialists. Therefore, this simple example underlines the real need for advanced methods and tools able to maintain automatically semantic annotations impacted by KOS evolution and/or changes in the annotated data or documents.

In this global context, this PhD project will provide methods and tools for supporting the (semi-)automatic maintenance of semantic annotations affected by KOS or the information evolution in order to keep annotations exploitable. The proposed framework will include different approaches to make annotations affected by changes evolve, either by modifying annotations directly or in an indirect way when annotations cannot directly be changed, e.g., in the case of encrypted EHRs. We are interested in answering the following research question within this project:

How can we (semi-)automatically maintain the validity of semantic annotations without re-annotation of documents after making changes in the underlying KOS?

2   State of the Art

Despite the heavy use of annotations especially in the life sciences, automatic or semi-automatic annotation maintenance has not been studied much so far. Traverso-Ribón et al. [15] developed the AnnEvol framework to compare two versions of a dataset (for instance, UnitProt-GOA and Swiss-Prot) and to verify the entities in the dataset(i) and dataset(i+1) that are similar and those which are different. The reported results suggest that annotations gradually evolve into groups of similar annotations, but also that evolution of the annotations not only depends on the organism of the protein (e.g. Homo Sapiens), but also on the type source of the annotation, i.e., Swiss-Prot and UniProt-GOA. Groß et al. [8] analysed how GO annotations evolved in the protein databases Ensembl and Swiss-Prot. They observed that a huge number of annotations changed over time, apparently triggered by changes in the underlying KOS as well as changes in the protein descriptions themselves. They also analysed the stability of annotations with respect to whether annotations have been determined manually or automatically. Park et al. [14] also analysed semantic inconsistencies in GO annotations and propose a correction method based on the hierarchical structure of the GO graph as well as tree structure of the NCBI taxonomy.

All these studies focus on the evolution of the popular GO and it is not clear whether GO-related observations about the stability of annotations and first ideas to adapt GO annotations are valid for other, larger KOS such as SNOMED CT or NCI Thesaurus and clinical annotation use cases. Annotation methods for clinical documents have only rarely been considered [3,13]. Existing approaches provide a general annotation pipeline based on smaller project-specific vocabularies but do not consider the evolution of large terminologies as well as the indirect maintenance of annotations. Previous results on the evolution of GO motivate the strong need for automatic annotation migration methods but these are still lacking for clinical use cases.

As mentioned in [3,16] the annotation changes and what exactly influenced the evolution of annotations or how to repair inconsistencies needs further research. In this direction, Luong et al. [12] investigate annotation evolution for the corporate Semantic Web and use a rule-based approach to detect and correct annotation inconsistencies. This approach focuses on expressive but small-sized ontologies and can hardly be applied to large biomedical KOS because the implemented reasoning techniques require the power of description logics to decide the validity of the annotations. Frost [6] proposes a novel algorithm for optimizing gene set annotations to best match the structure of specific empirical data sources. The proposed method uses Entropy Minimization over Variable Clusters (EMVC) to improve the annotations but, the authors highlight that EMVC only works in the gene set domain, thus other domains can not take advantage of this approach.

In summary, existing approaches for annotation evolution mostly handle only simple changes (e.g.: concept addition and deletion), consider only small domain ontologies and assume that the annotations are modifiable which is not always the case, e.g., for EHRs. They do thus not sufficiently address the problem of maintaining semantic annotations in biomedical and clinical use cases. In this PhD project, we aim at overcoming such shortcomings. We plan to build on our previous work such as the COnto-Diff [9] and DyKOSMap [4] tools but expect that significant changes become necessary, e.g., include background knowledge to support is-a and same-as match correspondences between KOS versions and to consider the diverse usage scenarios and semantics of annotations.

3   Proposed Approach

This PhD project aims to comprehensively address the requirements for annotation maintenance and to correct shortcomings of previous approaches. We will develop a formal framework for the (semi-)automatic maintenance of semantic annotations impacted by evolution in KOS or annotated documents in order to address the raised research question, see Sect. 1. We have four main objectives:

- **Providing a comprehensive annotation model:** In order to maintain the validity of annotations we will first investigate the current annotation models and propose useful parameters that must be included to cope with the annotation evolution problem. As mentioned in [5], annotation management and comparison is restricted to analyses of basic features and do not cover all cases of annotation evolution, e.g., "resurrection events". It means cases where an annotation is created in one release, later deleted, and then, after a lapse of one or more releases, a new annotation is created. Therefore, our annotation model needs to carefully consider different reasons of annotation evolution.
- **Characterizing the changes of KOS and/or annotated elements with fine level of granularity considering the semantics of change:** Annotation maintenance becomes necessary when the used KOS or the annotated data itself change. We thus need to determine all changes relevant for annotation maintenance in a detailed and fine-grained way. Differently from most of the correlated works, that just handle basic KOS changes, addition and removal of concepts, we will consider complex KOS changes [9]. It will allow us to have precise information about the KOS evolution to maintain the annotations up-to-date.
- **Developing methods for maintaining semantic annotations:** Invalid annotations need to be adapted to preserve their validity. A main aim of a (semi-)automatic annotation adaptation is to apply rules, use background knowledge and semantic change patterns to correct annotations no longer valid according to a current KOS version without re-annotating the documents. Differently from the correlated works, we will directly process the annotations if they are accessible, or in an indirect and ad-hoc way if annotations are not modifiable. Furthermore, the previous works do not reuse background knowledge that contains rich information about KOS evolution.
- **Validation and tool development:** Since the only gold standard corpus available containing annotations produced in different periods is UnitProt-GOA and Swiss-Prot from GO annotations, we plan to use a corpus of

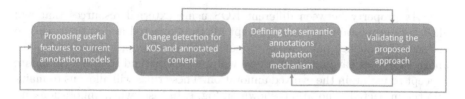

Fig. 2. Steps carried out to conduct the methodology. We are considering a iterative process with a feedback-based development cycle.

electronic health records and case report forms of clinical trials to generate annotations using other KOS as MeSH and SNOMED CT. Thus we will build a gold standard corpus to be reused in future evaluations.

4   Methodology

The adopted research methodology has the following steps, illustrated in Fig. 2:

- **Proposing useful features to current annotation models (see objective 1, Sect. 3):** We will use annotation tools such as GATE and BioPortal Annotator and/or ask human experts to generate the initial annotations. We therefore will analyse how (1) automatic annotation generation methods and (2) human experts realize the initial annotation of real-world objects. Automatically generated annotations might be of limited quality compared to manually verified ones. Therefore, our annotation model needs to carefully consider different reasons of annotation evolution, e.g., "resurrection events" see Sect. 3. As base model to represent the annotations, we will reuse the W3C Web Annotation Data Model[1].
- **Change detection for KOS and annotated content (see objective 2, Sect. 3):** Based on our previous annotation model, we will determine the requirements for identifying and describing the changes in the underlying KOS and annotated documents. In particular, we plan to develop a generalized matching algorithm to determine a semantically expressive mapping between concepts in successive KOS versions. Our algorithm will make use of the comprehensive concept context, e.g., $(c1 - superClass - c2)$ available in the KOS schema as well as valuable external knowledge like synonym sets (e.g. from UMLSs) and ontology changes provided by COnto-Diff [9]. Moreover, we will investigate the applicability of linguistic techniques [1].
- **Defining the semantic annotations adaptation mechanism (see objective 3, Sect. 3):** We will define rules to maintain the affected annotations. We will conceptualize and formalize heuristics: (i) to maintain annotations impacted by KOS changes, e.g., the evolution of annotations caused by a *split* concept; (ii) to use background knowledge to keep the annotations up-to-date, e.g., even a concept has a *ChangeAttributeValue*, we can find a

---

[1] W3C Web Annotation Data Model.

sameAs property between different KOS using external resources that will allow us to maintain the same concept for the impacted annotations; (iii) to use semantic change patterns as support method [4], e.g., in cases where the above heuristics have not worked, we will use this method to find a candidate concept to maintain the evolved annotation. These rules will allow us to maintain the impacted annotations shown on Fig. 1. In cases when annotations are not modifiable, we plan to expand queries with updated annotations.

- **Validating the proposed approach (see objective 4, Sect. 3):** The proposed solutions will be evaluated using the data of the TREC document corpora. Moreover, we plan to evaluate our methods on real applications using the EHR data from the eSanté platform as well as CRFs from the LIFE project[2]. In particular, we will analyse the evolution of KOS like SNOMED CT, MeSH, NCIT and ICD-9-CM and associated annotations. The results will be produced regarding the precision and recall from the semantic annotation adaptation mechanism.

5   Results

In our first investigations we wanted to observe how KOS evolution impacts on the validity of semantic annotations. To this end, we produced more than 64 million annotations using GATE and BioPortal annotation tools and the ICD-9-CM and MeSH from UMLS, periods 2004AA to 2016AA (see Table 1). Our primary results in [2] show different amounts of annotations regardless the used KOS and a clear correlation between changes in KOS and the validity of the annotations over time.

We also observed the necessity to improve the used annotation data model (see Sect. 4). The current W3C model (Fig. 3a) does not allow us to create a link chain between the evolved annotations. Therefore we developed a new feature called "*evolved to*" (see Fig. 3b). This feature allows us to access the previous versions of an annotation, instead of using the Memento protocol[3] or the feature "*cached field*" in the W3C model for example.

The various experimentations also allowed us to design the global framework for maintaining semantic annotations. It implements a 4-step methodology that

Table 1. Amount of annotations produced with the tools: BioPortal and GATE

Annotator	KOS	Versions	Amount
GATE	ICD-9-CM	UMLS: 2004AA to 2016AA	2,675,170
BioPortal	ICD-9-CM	UMLS: 2004AA to 2016AA	2,200,655
GATE	MeSH	UMLS: 2004AA to 2016AA	32,816,092
BioPortal	MeSH	UMLS: 2004AA to 2016AA	26,879,336

---

[2] LIFE project.
[3] Memento protocol.

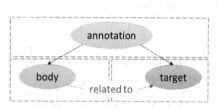

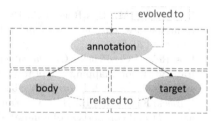

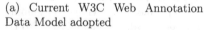

(a) Current W3C Web Annotation Data Model adopted

(b) Proposed feature to W3C Web Annotation Data Model

Fig. 3. Modifications in the W3C Web Annotation Data Model after analysing our primary results and methods to access the changed annotations

relies on several techniques including logic rules and background knowledge to adapt invalid annotations.

6   Conclusion

This project focuses on knowledge engineering and semantic interoperability within information systems for medical or biomedical domains. Tools which allow sharing and exploiting the huge amount of health-related data are highly demanded by the medical community. Since this work just completed one year we plan as future steps to finish the rules used to correct the impacted annotations and evaluate the first results using the automatic dataset produced with GATE and BioPortal. Furthermore, we are establishing relationships with domain specialists to access their manual annotations that will allow us to improve the created rules for maintaining the annotations.

Our mid-term goal is to develop the framework using a rule-based approach, background knowledge and semantic change patterns to correct the impacted annotations. We expect as main aim to develop a semi-automatic approach using machine learning to correct the evolved annotations.

Acknowledgment. The author would like to thank both National Research Fund (FNR), who supported this work, and his advisors: Cédric Pruski, Marcos Da Silveira and Chantal Reynaud-Delaître.

References

1. Arnold, P., Rahm, E.: Semantic enrichment of ontology mappings: a linguistic-based approach. In: Catania, B., Guerrini, G., Pokorný, J. (eds.) ADBIS 2013. LNCS, vol. 8133, pp. 42–55. Springer, Heidelberg (2013). doi:10.1007/978-3-642-40683-6_4
2. Cardoso, S.D., Pruski, C., Silveira, M., Lin, Y.-C., Groß, A., Rahm, E., Reynaud-Delaître, C.: Leveraging the impact of ontology evolution on semantic annotations. In: Blomqvist, E., Ciancarini, P., Poggi, F., Vitali, F. (eds.) EKAW 2016. LNCS, vol. 10024, pp. 68–82. Springer, Cham (2016). doi:10.1007/978-3-319-49004-5_5

3. Da Silveira, M., Dos Reis, J.C., Pruski, C.: Management of dynamic biomedical terminologies: current status and future challenges. Yearb. Med. Inform. **10**(1), 125–133 (2015)

4. Dos Reis, J.C., Pruski, C., Da Silveira, M., Reynaud-Delaître, C.: Dykosmap: a framework for mapping adaptation between biomedical knowledge organization systems. J. Biomed. Inform. **55**, 153–173 (2015)

5. Eilbeck, K., Moore, B., Holt, C., Yandell, M.: Quantitative measures for the management and comparison of annotated genomes. BMC Bioinform. **10**(1), 67 (2009)

6. Frost, H.R., Moore, J.H.: Optimization of gene set annotations via entropy minimization over variable clusters (emvc). Bioinformatics **30**(12), 1698–1706 (2014). (Oxford, England)

7. Funk, C., Baumgartner, W., Garcia, B., Roeder, C., Bada, M., Cohen, K.B., Hunter, L.E., Verspoor, K.: Large-scale biomedical concept recognition: an evaluation of current automatic annotators and their parameters. BMC Bioinform. **15**(1), 1–29 (2014)

8. Groß, A., Hartung, M., Prüfer, K., Kelso, J., Rahm, E.: Impact of ontology evolution on functional analyses. Bioinformatics **28**(20), 2671–2677 (2012)

9. Hartung, M., Gross, A., Rahm, E.: Conto-diff: generation of complex evolution mappings for life science ontologies. J. Biomed. Inform. **46**, 15–32 (2013)

10. Hodge, G.: Systems of knowledge organization for digital libraries: beyond traditional authority files. Reports - Descriptive (2000)

11. Lowe, H.J., Barnett, G.O.: Understanding and using the medical subject headings (mesh) vocabulary to perform literature searches. JAMA **271**(14), 1103–1108 (1994)

12. Luong, P.H., Dieng-Kuntz, R.: A rule-based approach for semantic annotation evolution in the CoSWEM system. In: Koné, M., Lemire, D. (eds.) Canadian Semantic Web. Semantic Web and Beyond, pp. 103–120. Springer, Heidelberg (2006)

13. Mowery, D.L., Jordan, P.W., Wiebe, J., Harkema, H., Chapman, W.W.: Semantic annotation of clinical events for generating a problem list. In: AMIA Annual Symposium proceedings/AMIA Symposium, AMIA Symposium 2013, pp. 1032–1041 (2013)

14. Park, Y.R., Kim, J.J.H., Lee, H.W., Yoon, Y.J.: GOChase-II: correcting semantic inconsistencies from gene ontology-based annotations for gene products. BMC Bioinform. **12**(Suppl 1), S40 (2011)

15. Traverso-Ribón, I., Vidal, M.-E., Palma, G.: AnnEvol: an evolutionary framework to description ontology-based annotations. In: Ashish, N., Ambite, J.-L. (eds.) DILS 2015. LNCS, vol. 9162, pp. 87–103. Springer, Cham (2015). doi:10.1007/978-3-319-21843-4_7

16. Uren, V., Cimiano, P., Iria, J., Handschuh, S., Vargas-Vera, M., Motta, E., Ciravegna, F.: Semantic annotation for knowledge management: requirements and a survey of the state of the art. Web Semant. Sci. Serv. Agents World Wide Web **4**(1), 14–28 (2006)

Photo Archives in Linked Open Data – The Federico Zeri's Archive Case Study

Marilena Daquino[✉]

Department of Classical Philology and Italian Studies,
University of Bologna, Bologna, Italy
marilena.daquino2@unibo.it

Abstract. Art historical photo archives that want to expose their data in Linked Open Data need to rely on shareable models. Merging possibly contradictory information may affect data reliability. In this paper are introduced two ontologies, i.e. F Entry Ontology and OA Entry Ontology, which provide a complete description of items related to Photography and Arts domains and address the description of questionable information provided by different institutions. A preliminary analysis of the Zeri's photo archive was performed for guiding the creation of the ontologies and the mapping of all the partners' metadata schemas.

Keywords: Linked Open Data · Ontology development · Photo archives

1 Introduction

Art historical photo archives collect and describe heterogeneous documents reproducing works of art and the chronology of authorship attributions for the sake of comparison in historical research [2]. Although such information is meant to be preserved over the long term, data of photo archives might evolve over time. New sources, criteria, and the integration of lacking information provided by similar data sources can affect the validity of an attribution.

New approaches should be investigated in the knowledge representation of photo archives. The annotation of provenance, methods and criteria, and the contextualization in time and space of questionable information, would be the key to ensure the optimal data reuse. Managing properly this scenario will prevent duplication of intellectual efforts when merging knowledge bases or when creating a new one. Capturing and sharing the implicit, subjective, knowledge that libraries and archives have, aside the explicit and documented knowledge, will improve knowledge transfer and access to other cultural institutions [1], avoiding inconsistencies when integrating data sources.

The application of Linked Open Data and Semantic Web technologies is considered a promising approach to address these issues. In the last years, archives converted their cataloguing standards into a large number of RDFSchemas and ontologies, but different approaches still affect a (uniform) representation of the domain, and don't provide an exhaustive mapping of all the existent descriptive

© Springer International Publishing AG 2017
P. Ciancarini et al. (Eds.): EKAW 2016 Satellite Events, LNAI 10180, pp. 219–223, 2017.
DOI: 10.1007/978-3-319-58694-6_34

needs (see Sect. 2). In fact, none of the existing and widely used models satisfy the whole description of a single photo archive.

To achieve these goals in the context of photo archives, the *International Consortium of Photo Archives (PHAROS)* was created in 2014 [13]. The purpose is the creation of an environment – mainly aimed at researchers, in particular historians – where to expose images and related metadata provided by fourteen of the most relevant photo archives in Europe and United States. The first concrete step in this direction was lead by the Federico Zeri Foundation of the University of Bologna, which is in charge to map its own cataloguing standards to RDF, in order to guide the definition of a common model.

The contribution to the project here presented regards the mapping of metadata content standards used by the Zeri's photo archive to define two models for describing photographs and artworks related metadata and the aforementioned issues related to questionable information. The aim is to properly describe all the information of importance to final users, e.g. the detection of the most authoritative attribution by means of semi-automatic processes and rankings.

2   Related Works

Modelling heterogeneous information provided by photo archives involves different domains, i.e., museums, archives, libraries, arts, photography and history. While current models in library and publishing domains mainly focus on a FRBR-compliant [7] description of their objects (e.g., RDA [6], SPAR ontologies [12]), museums prefer an event-driven approach (e.g. CIDOC-CRM [8], FRBRoo [9]). Archives, by contrast, only address the description of the hierarchical structure of the archive rather than the description of the documents (Reload [14], SAN [15]). Models trying to match similarities between these domains are still lacking. The photography field still lack of a formal representation as well.

In 2014, the Federico Zeri Foundation started the mapping to RDF of a sample of its cataloguing rules for describing photographs, called *Scheda F* (i.e., F Entry[1]), which resulted in the development of the F Entry Ontology (FEO), http://www.essepuntato.it/2014/03/fentry.

This research extends the former one [5], by means of the realization of a new, specular ontology, i.e., the OA Entry Ontology, available at http://purl. org/emmedi/oaentry. It aims at the mapping of *Scheda OA* (i.e., OA Entry[2]), the cataloguing standard adopted for describing artworks. In particular, it includes an *ad hoc* developed task ontology, HiCO (available at http://purl.org/emmedi/ hico), for describing the interpretative process underlying questionable information. The PROV Ontology [10]with extensions was preferred for the development of HiCO, as detailed in [4].

---

[1] http://www.iccd.beniculturali.it/index.php?it/473/standard-catalografici/ Standard/10.

[2] http://www.iccd.beniculturali.it/index.php?it/473/standard-catalografici/ Standard/29.

3   A Modular and Iterative Approach to Expose Linked Open Data of Photo Archives

A special attention is here given to the subjective and implicit knowledge "owned" by photo archives. Authority files and controlled vocabularies don't cover motivations, methodologies applied in the cataloguing process. Defining them in order to describe the extent in which an information is considered authoritative – e.g. supported by bibliographic citations, material evidences, or authoritative opinions – aims at a change of librarians and archivists' perspective on their data, by creating data that can be enriched or contradicted over time.

The proposal is to map cataloguing rules used by PHAROS partners in an iterative, incremental revision of the F Entry and OA Entry ontologies. Completeness of the resulting mapping is ensured, since all the really used metadata are described in the final models, and data integration will be reached without lost of information. As further described in Sect. 4, models are iteratively tested on existent data. The quality of the mapping can be assessed by considering consistency of both models and data, aside the reusability of the mapping itself by other kind of cultural institutions. Finally, trustworthiness of cataloguers that evaluate the mapping ensures its correctness.

At now, several domain and task ontologies are part of the two modular ones, i.e., the F Entry Ontology and OA Entry Ontology. Each model covers the description of a specific issue: a FRBR-compliant description of the objects (FaBiO, http://purl.org/spar/fabio); the life cycle of the objects (CIDOC-CRM) and the involved roles (PRO, http://purl.org/spar/pro); relations to other cultural objects (CiTO, http://purl.org/spar/cito and PROV); and finally, provenance of questionable information (HiCO).

4   Methodology

To achieve the above described goals, several tasks are required, including:

1. the study of relevant national and international cataloguing standards and the analysis of state-of-the-art ontologies for cultural heritage domain;
2. the conversion into RDF according to developed models of data provided for testing purposes;
3. the definition of a conceptual framework for making further assertions on the validity of contradictory statements, considering the need to enhance low-quality metadata with information about provenance, time and space contextualization, methods and criteria.

By starting from one of the most complete use cases, i.e. the Zeri Photo Archive, all the main descriptive issues are addressed at first (points 1–3). The SAMOD methodology [11] for ontology development was preferred as it is based on well-known ontology development methodologies and is particularly data-centric. Indeed, it requires to test models on data iteratively.

When a new photo archive is taken into account, refinement of models and data integration proceed together. They will be integrated in order to show how contradictory information may coexist and how can enrich the knowledge base. Data and models will be validated by a group of chosen users, including cataloguers and researchers. They will check consistency and completeness of data and will provide feedback useful to refine the set of shareable criteria in use for defining authoritative attributions.

5   The Zeri Photo Archive Case Study

At the end of the first year, a preliminary analysis was performed on the subset of cataloguing standards really used by the Zeri catalog. About 120 fields out of more than 300 provided by *Scheda F* and about 100 fields out of 280 provided by *Scheda OA* have been mapped to the aforementioned ontologies.

Current outcomes of the project, further detailed in [3], are:

- a first version of the OA Entry Ontology and a revision of the F Entry Ontology;
- two mapping documents from OA/F content standards to RDF, respectively, *OAEntry to RDF* (https://dx.doi.org/10.6084/m9.figshare.3175057) and *F Entry to RDF* (https://dx.doi.org/10.6084/m9.figshare.3175273);
- the RDF dataset of the Zeri's photo archive (https://w3id.org/zericatalog).

At now, data published describe about 30.000 photographs, 19.000 artworks, related archival documents, and artists, which are represented by about 11 million RDF statements.

6   Conclusions and Future Work

Future outcomes of the project will include the revision of the ontologies, considering few aspects deemed secondary at this first stage of development and the analysis of the other PHAROS partners metadata schemas will be performed. RDF data created for testing models will be likely published as Linked Open Data too, although this is out of the scope of this research.

The aim is to perform an analysis on the data with regard to the reviewed criteria adopted by cataloguing institutions, art critics and historians and to test semi-automatic mechanisms for detecting authoritative authorship attributions, e.g. using SWRL rules (https://www.w3.org/Submission/SWRL/). Such improvements will lead to an enhancement of the quality of data published by photo archives, especially with regard to how they reached possible contradictory information.

Acknowledgments. I would like to express my grateful thanks to my supervisor Dr. Francesca Tomasi (University of Bologna) for her ongoing support in writing this thesis.

References

1. Bultrini, L., McCallum, S., Newman, W., et al.: Knowledge Management in Libraries and Organizations. De Gruyter Saur, Berlin, Boston (2015)
2. Caraffa, C.: Photo Archives and the Photographic Memory of Art History. Deutscher Kunstverlag, Berlin (2011)
3. Daquino, M., et al. Enhancing semantic expressivity in the cultural heritage domain: exposing the Zeri Photo Archive as Linked Open Data. arXiv preprint arXiv:1605.01188 (2016)
4. Daquino, M., Tomasi, F.: Historical Context (HiCO): a conceptual model for describing context information of cultural heritage objects. In: Garoufallou, E., Hartley, R.J., Gaitanou, P. (eds.) Metadata and Semantics Research. Communication in Computer and Information Science, vol. 544, pp. 424–436. Springer, Cham (2015)
5. Gonano, C.M., Mambelli, F., Peroni, S., Tomasi, F., Vitali, F.: Zeri e LODE: Extracting the Zeri photo archive to linked open data: formalizing the conceptual model. In: Proceedings of the 2014 IEEE/ACM Joint Conference on Digital Libraries (JCDL 2014), pp. 289–298. IEEE, Washington (2014)
6. Hillmann, D., et al.: RDA vocabularies: process, outcome, use. In: D-Lib magazine, vol. 16(1). Corporation for National Research Initiatives (2010)
7. IFLA Study Group on the Functional Requirements for Bibliographic Records and Standing Committee of the IFLA Section on Cataloguing: Functional Requirements for Bibliographic Records. Final Report. K.G. Saur, Berlin, Boston (2013)
8. Le Boeuf, P., Doerr, M., Ore, C.E., Stead, S.: Definition of the CIDOC Conceptual Reference Model (2015). http://www.cidoc-crm.org/docs/cidoc_crm_version_6.2.1.pdf
9. Le Boeuf, P.: A strange model named FRBRoo. Cataloging Classif. Q. **50**(5–7), 422–438 (2012). Taylor & Francis
10. Lebo, T., Sahoo, S., McGuinness, D.: PROV-O: The PROV Ontology. W3C Recommendation, 30 April 2013. http://www.w3.org/TR/prov-o/
11. Peroni, S.: SAMOD: an agile methodology for the development of ontologies. figshare. http://dx.doi.org/10.6084/m9.figshare.3189769
12. Peroni, S.: The semantic publishing and referencing ontologies. Semantic Web Technologies and Legal Scholarly Publishing. LGTS, vol. 15, pp. 121–193. Springer, Cham (2014). doi:10.1007/978-3-319-04777-5_5
13. Reist, I., Farneth, D., Stein, R.S., Weda, R.: An introduction to PHAROS: aggregating free access to 31 million digitized images and counting. In: Speech at: CIDOC 2015, New Delhi, 5–9 September (2015). http://network.icom.museum/fileadmin/user_upload/minisites/cidoc/BoardMeetings/CIDOC_PHAROS_Farneth-Stein-Weda_1.pdf
14. Repository for Linked Open Archival Data (ReLoad). http://labs.regesta.com/progettoReload/
15. SAN Ontology. Documentation. http://dati.san.beniculturali.it/SAN/

Dealing with Velocity and Variety
in the Acquisition of Heterogeneous Sensor Data

Luca Ferrari[✉]

Department of Computer Science, University of Milano, Milan, Italy
lferrari@di.unimi.it

Abstract. ETL (Extraction-Transform-Load) tools, traditionally developed to operate offline, need to be enhanced to deal with various, fast, big and fresh data and be executed on the edge of the network during the acquisition process. In this dissertation we wish to develop facilities that from one side make easy, scalable and controllable the development of data acquisition plans that can be executed on the edge of the network during loading and transmission. From the other side, we wish to deal with the variety of the data and verify when the developed data acquisition plans adhere to the common semantics adopted in the Domain Ontology. These facilities are included in StreamLoader, a web application tailored for the specification and monitoring of sensor data acquisition plans.

Keywords: ETL · IoT · IoT Ontology · Sensors · In-network computing

1 Introduction

The Internet of Things (IoT) is an all-encompassing and ubiquitous network of devices that helps coordination and communication between heterogeneous devices, able to detect physical phenomena (like, temperature, humidity, wind), and social events (like, twitter data, traffic information). Data produced by these sensors are heterogeneous in structures (different types), in spatial and/or temporal granularities, in thematics.

The challenge is to generate different services, in different domain, able to extract valuable information from these different devices/sensors, transform them to fit the analytical needs, and easily load them into a DW for subsequent analysis. These services should be applied during data acquisition and bound with reactive capabilities in order to properly identify the relevant streams when abnormal events occur and undertake the proper actions. Finally, the specification and actuation of these services should be efficiently performed on-line and on fresh and timely data to properly handling big real-time data streams.

In order to address these issues we are working on *StreamLoader*, a Web application for the specification of conceptual ETL dataflows on heterogeneous sensors to be applied during data stream acquisition on a programmable network. StreamLoader is equipped with an interactive and easy-to-use environment

© Springer International Publishing AG 2017
P. Ciancarini et al. (Eds.): EKAW 2016 Satellite Events, LNAI 10180, pp. 224–229, 2017.
DOI: 10.1007/978-3-319-58694-6_35

that supports the user in charge of handling events to discover the sensors useful in a given situation, specify the adequate dataflow for extracting, filtering, integrating, (eventually) storing, and analyze the data coming from the identified sensors, optimize the schedule for the execution of the dataflow and visualize the results. During the execution, the user can eventually modify on the fly the dataflow and his modifications are automatically applied without stopping and restarting the data acquisition plan. In this environment we move the computation from a central processor to the edge of the network, where data are actually produced. This allows us to filter out non-relevant data, compress/aggregate data, and automatically generate metadata supporting a rich data description.

In the reminder, Sect. 2 discusses related work. Section 3 presents a motivating example and the description of the overall architecture. Section 4 introduces the key ingredients for the specification of the data acquisition plan: the STT data model, the Domain Ontology, and the services that can be composed. Section 5 deals with the characteristics of the StreamLoader system. Section 6 concludes.

2   Related Work

In this section we discuss sensor data acquisition and processing techniques. Then, we deal with the facilities developed for the specification and execution of ETL operations. We conclude with the on-going efforts to define ontologies for the representation of semantic sensor networks.

Sensor Data Acquisition Techniques. Sensor data needs to be processed in order to identify useful events in a given domain. Two main processing strategies can be devised. The *Data warehousing approach* involves shipping raw sensor readings from the sensor network to a central repository for subsequent analysis. This solution is appropriate when all data needs to be kept and there is sufficient energy to transmit the data outside the sensor network and/or storage is possible. Recently a number of open source data streaming tools has been proposed (such as Apache S4 [6] and Storm [4]) that apply the *Map-Reduce Paradigm* to realize scalable and efficient solutions. By contrast *in-network processing* strategy is critical when: data quality needs to be quickly analyzed; energy saving is a concern; and, there is the need to save communication costs and reduce the latency. ETL services are applied during the transmission of the data from the sensors to the collecting nodes and different optimizations techniques can be adopted (like Dynamic Query re-Optimization [7] and Progressive Optimization [8]).

ETL Visual Facilities. Many systems have been proposed for the specification and actuation of ETL operations. Most of them differ from the kinds of data they can handle (structured and semi-structured). Commercial systems like Talend Studio (www.talend.com), Waylay.io (www.waylay.io) offer GUIs for designing workflows/dataflows as graphs of connected nodes representing tasks and data-sources. However, all of them are quite complex to use, seldom provide web GUIs for designing and monitoring dataflows and are not integrated in a single tool. This limits their use.

IoT Ontologies. Nowadays, many on-going efforts are devoted to the definition of ontologies and the design of frameworks to apply semantic Web technologies to sensor networks. The Semantic Sensor Web (SSW) proposes to describe sensor data with semantic metadata [9] able to specify the capabilities of sensors, the measurement processes and the resultant observations. W3C Semantic Sensor Networks (SSN) Incubator Group [3] worked on developing an ontology for describing sensors. The core concepts and relations of the SSN ontology [3] concern the description of sensors, features, properties, observations, and systems. Then, measuring capabilities, operating and survival restrictions, and deployments were added in turn. The IoT-Lite ontology [2] has been introduced starting from the SSN ontology.

3   Motivating Example and Architecture

Suppose that many sensors are disseminated in the area of Milano that can be exploited for the computation of the Apparent Temperature[1] (AC) when the number of tweets per hours about hot temperature is greater than 20. However, the sensors return the events using different formats, and with different spatio-temporal granularities. Sensors of type T_1 allow to gather the temperatures every 10 min in Celsius degree in JSON format. The sensors of type H_1 allow to generate humidity events every 30 min with the XML format (but no geo-spatial location is provided). Finally, sensors of type TW_1 generate every minutes the list and the total number of tweets that are exchanged in a given zone of the city in the CSV format. The information needed for computing AC is thus present in these sensors (and in their metadata) but need to be acquired, normalized, integrated and elaborated before being ready for the computation. For example the data of sensors of type H_1 has to be enriched with information about their spatial, temporal and thematic dimensions. With this goal, a data acquisition plan should be defined that can be applied to the streams of events that are generated by the different sensors in order to produce the information required for the computation of AC. First, the schema of each sensor needs to be mapped to an internal data model in which, when available, the temporal, spatial and thematic dimensions are pointed out. This guarantees the adoption of a common model within our system (tough at this level we cannot guarantee the adoption of the same semantics). Then, some services are composed for filtering, combining, convert to the same granularity and enhancing the produced events.

---

[1] Apparent temperature is a value of temperature adjusted with the level of humidity.

4   Proposed Approach

In this section we detail our sensor data model according to the STT dimensions. Then, we detail the characteristics of the Domain Ontology and provide a brief description of the proposed data acquisition services.

STT Model. In our model both simple (like integer, float, string) and structured (records and lists) values can be represented along the multi-granular spatio-temporal-thematics dimensions. The temporal dimension relies on the Gregorian Calendar and allows to represent instant of time at different granularities. The spatial dimension allows to represent geometric objects represented in one or two dimensions (like points, lines, and regions). Examples of temporal granularity include second, minute, day whereas, meter, kilometer, feet, yard, zone and city are examples of spatial granularities. The Thematic dimension represents the type of events that is generated by sensors. Examples of thematics are temperature, humidity, wind speed, etc. Thus in our model, an *event* is an instance of a thematic associated with a spatio-temporal granularity.

Domain Ontology. The STT data model so far presented allows one to produce events whose correctness is left to the user in charge of its creation. To support the user in this activity, a Domain Ontology is included within our system. The Domain Ontology is an extension of the SSN ontology [3] generated by the domain experts by giving a precise meaning of the spatio-temporal-thematic dimensions that are used in their context. The purpose of the Ontology is to guarantee that the properties specified for a given concept actually occur in the data produced by the sensors and that the events that are generated at the end of the dataflow are compliant with spatial-temporal and thematic data model that we use.

Data Acquisition Services. A set of services/operators has been defined for processing and combining the streams produced by the sensors. We can distinguish two kind of operators: *non-blocking operators* and *blocking operators*. The former are applied of each single event and thus do not require to maintain caches. The latter requires to maintain a caches of events for a temporal interval t. The operation is applied at the end of the interval with the cached events. A detailed description of the operators is reported in [5].

The combination of services in a dataflow should follow some rules in order the have *sound* and *consistent* data acquisition plans. This is relevant for obtaining execution plans that can be computed without errors in the underlying network and that adhere to our Ontology. Specifically, a data plan is *sound* if for each applied service the number of input streams is equal to the number of expected input streams; the parameters are specified; and, the applicability conditions are verified. Moreover, a sound data plan is *consistent* when the schema of the output stream of generated events is consistent w.r.t. the DO.

5   The StreamLoader System

By considering the motivating example discussed in Sect. 3, in this section we provide some details of the StreamLoader GUI interface and how the dataflow is executed at network level.

A web-based application has been developed. This interface allows users to create dataflows by dragging and dropping operators on a canvas. Once the dataflow is ready, the plan is translated into a declarative language (DSN) and logs are monitored during the execution.

SCN [1] allows us to execute the DataFlow over a Software Defined Networking at in-network level. SCN automatically configure the network and reconfigure the paths while congestions or node failures occur.

6   Conclusions

In this paper we described the Ph.D. activities we are carrying on to develop a solution for the acquisition of heterogeneous data streams generated by sensors distributed on a network. Preliminary results have been reported in [5]. The idea is to capture the context in which data has been generated and, by applying specific services, to filter out non-relevant data, compress/aggregate data, and automatically generate metadata supporting a rich data description. These operations are carried out by a Web application, named StreamLoader, which aim is to provide users with an user-friendly environment for supporting the design and execution of the data acquisition specification. The result of this activity is a dataflow that can be correctly activated at network level and so translated in services of a programmable network by means of the SCN facilities. Thanks to the adoption of a DO we can help domain experts in discovering and adding new operators and for facilitating the semantic integration.

As future work we will first conduct experiments, in order to evaluate the performance of the whole system and to compare the performance of the in-network approach w.r.t. the data warehousing approach. In this last direction, we wish to compare the in-network approach we are developing with an implementation of the ETL operations in Apache Spark [10].

References

1. Dong, M., Kimata, T., Zettsu, K.: Service-controlled networking: dynamic in-network data fusion for heterogeneous sensor networks. In: IEEE International Symposium on Reliable Distributed Systems Workshops (SRDSW), pp. 94–99 (2014)
2. Bermudez-Edo, M., et al.: IoT-Lite ontology (2015)
3. W3C Semantic Sensor Network Group: Semantic sensor network ontology (2005)
4. Ankit, J., et al.: Learning Storm. Packt Publishing, Birmingham (2014)
5. Mesiti, M., et al.: StreamLoader: An event-driven ETL system for the on-line processing of heterogeneous sensor data. In: Proceedings of International Conference on Extending Database Technology, pp. 628–631 (2016)

6. Neumeyer, L., et al.: S4: distributed stream computing platform. In: International Workshop on Data Mining ICDMW, pp. 170–177 (2010)
7. Kabra, N., et al.: Efficient mid-query re-optimization of sub-optimal query execution plans. SIGMOD Rec. **27**(2), 106–117 (1998)
8. Markl, V., et al.: Robust query processing through progressive optimization. In: Proceedings of International Conferences on Management of Data, SIGMOD, pp. 659–670 (2004)
9. Sheth, A., et al.: Semantic sensor web. IEEE Internet Comput. **12**(4), 78–83 (2008)
10. Karau, H., et al.: Learning Spark: Lightning-Fast Big Data Analysis. O'Reilly Media, Beijing (2015)

Semantic Data Integration
for Industry 4.0 Standards

Irlán Grangel-González[✉]

Enterprise Information Systems (EIS) Group,
University of Bonn and Fraunhofer (IAIS), Sankt Augustin, Germany
grangel@cs.uni-bonn.de

Abstract. Industry 4.0 initiatives have fostered the definition of different standards, e.g., AutomationML or OPC UA, allowing for the specification of industrial objects and for machine-to-machine communication in Smart Factories. Albeit facilitating interoperability at different steps of the production life-cycle, the information models generated from these standards are not semantically defined, making the semantic data integration a challenging problem. We tackle the problems of integrating data from documents specified either using the same or different Industry 4.0 standards, and propose a rule-based framework that combines deductive databases and Semantic Web technologies to effectively solve these problems. As a proof-of-concept, we have developed a Datalog-based representation for AutomationML documents, and a set of rules for identifying semantic heterogeneity problems among these documents. We have empirically evaluated our proposed framework against several benchmarks and the initial results suggest that exploiting deductive and Semantic Web techniques allows for increasing scalability, efficiency, and coherence of models for Industry 4.0 manufacturing environments.

1 Problem Statement

The Industry 4.0 vision aims at creating Smart Factories by combining the Internet of Things, Internet of Services, and Cyber-Physical Systems. To support this vision, Industry 4.0 communities have fostered the definition of standards such as AutomationML (IEC 62424) and OPC UA (IEC 62541). AutomationML is one of the core standards of Industry 4.0 for exchanging plant engineering information as specified by [1,7,16,18]. AutomationML can describe plant components and their sub-components from different views such as mechanical, electrical, or software. OPC UA [4] also allows for the description of the production life-cycle in Smart Factories, but contrary to AutomationML which describes characteristics of plant components, OPC UA models machine-to-machine communication. Although Industry 4.0 standards provide the basis for data exchange in a Smart Factory, information models of these standards require being aligned to facilitate the merging of a virtual process with real production life-cycles.

Smart Factories along with Cyber-Physical concepts impose new challenges to traditional approaches of data integration. A new generation of data-centric

© Springer International Publishing AG 2017
P. Ciancarini et al. (Eds.): EKAW 2016 Satellite Events, LNAI 10180, pp. 230–237, 2017.
DOI: 10.1007/978-3-319-58694-6_36

systems need to be developed and integrated, where data meaning, as well as data variety, veracity, and adaptivity must be managed [12]. In this context, achieving semantic data interoperability techniques suitable for these data properties, is of paramount importance for making the Industry 4.0 vision a reality.

With the aim of assessing the integration of Industry 4.0 standards, Biffl et al. [1] and Kovalenko and Euzenat [9] have characterized seven semantic heterogeneity issues among different views of an artifact. (**M1**) Value processing–same properties are not modeled equally, e.g., using different datatypes; (**M2**) Granularity–same objects are modeled at different levels of detail; (**M3**) Schematic differences–differences in the way how semantics is represented for the same object; (**M4**) Conditional mappings–relations between entities exist only if certain conditions occur; (**M5**) Bidirectional mappings–relations between entities have to be defined bidirectionally; (**M6**) Grouping and aggregation–different semantic modeling criteria are applied to group elements for the same object; and (**M7**) Restrictions on values–mandatory values for properties in the object that have to be handled in the mapping process. These semantic heterogeneity issues may occur between documents defined in the same or different standards, i.e., interoperability can be *intra-* or *inter-standard*.

Industry 4.0 standards and initiatives for identifying heterogeneity issues evidence the success of the Industry 4.0 movements. However, integration of standards are still conducted manually [5,17], negatively affecting the effectiveness of the production process. This doctoral work attempts to achieve two research goals to solve these problems. **RG1**: Addressing *intra-standard interoperability* among multiple pieces described in one standard, e.g., AutomationML. **RG2**: Assessing *inter-standard interoperability* in documents specified in different standards, e.g., documents in AutomationML and OPC UA.

To accomplish our research goals **RG1** and **RG2**, we propose a rule-based system that combines Deductive databases and Semantic Web technologies to effectively integrate documents specified in Industry 4.0 Standards.

2   State of the Art

Related work is divided into two sub-sections: (1) approaches for solving *intra-standard interoperability* with the focus in AutomationML; and (2) approaches for addressing *inter-standard interoperability* on AutomationML and OPC UA.

Intra-standard interoperability issues for AutomationML. In the literature, many different approaches are proposed for integrating AutomationML documents. In [17], a tool to map two AutomationML files is presented. It allows for the integration of AutomationML documents, their respective descriptions, and the modified parts of one file into the other. Further, a mapping algorithm for AutomationML files is presented. Nevertheless, the process of mapping is performed manually. Himmler [8] presents a framework to create standardized application interfaces in plant engineering based on AutomationML. The work provides a function-based based standardization framework for the plant engineering domain. Persson et al. [13] utilize an RDF-based approach to integrate robotized production information modeled AutomationML. Kovalenko et al. [10] explore how AutomationML can be represented by means of Model-Driven Engineering and the Semantic Web. A small part of an AutomationML ontology is developed, based on the main concepts of the language. Also, the use of rules for consistency checking is proposed, using the Semantic Web Rule Language (SWRL), but no explicit definition of the role of Semantic Web technologies on the integration problem is presented. The *AutomationML Analyzer* [14] is an online tool to browse, query and analyse different AutomationML data by means of Semantic Web technologies. A conceptual design to overcome integration problems in AutomationML is described. All these approaches have the potential to solve specific integration problems for AutomationML. However, they solve rather isolated problems, and a general method capable to automatically integrate AutomationML information from different perspectives is not provided.

Inter-standard interoperability issues for AutomationML and OPC UA. The current integration approach is performed by analyzing common information elements and manually describing the AutomationML objects in the OPC UA language [5,7,15]. In addition, in these works are documented the mappings between the common objects, i.e., their common structures and datatypes. To date, the possibility to semi-automatically integrate the information models of these two languages exploiting their semantic descriptions is still missing.

3   Proposed Approach

We propose ALLIGATOR [6], a rule-based framework for the semantic integration of Industry 4.0 standard documents. ALLIGATOR relies on Datalog and RDF to *accurately* represent the knowledge that characterizes different types of semantic heterogeneity for these documents. Further, by utilizing Datalog as well as Semantic Technologies, ALLIGATOR will be able to provide explanations for the alignments that occurred among the elements of the Industry 4.0 standard documents.

Figure 1 depicts the main components of the proposed solution. In the following we describe each component. The input for ALLIGATOR are Industry 4.0 standard documents. Next, these documents are translated into a canonical representation of a production life-cycle named the ALLIGATOR data model. RDF, RDFS, and domain-specific vocabularies are used to represent the core concepts of the ALLIGATOR data model. Further, standards documents are modeled as facts in an extensional database (EDB) of a Datalog program. Datalog rules comprise an intensional database (IDB), and state different types of semantic heterogeneity and *intra-* and *inter- standard interoperability* problems. The ALLIGATOR *Deductive System Engine* performs a bottom-up evaluation of ALLIGATOR Datalog programs following a semi-naïve algorithm that stops when the least fixed-point is reached [2]. The intensional predicates inferred in the evaluation of a ALLIGATOR Datalog program correspond to problems of semantic heterogeneity, and *intra-* and *inter-standard interoperability*. AutomationML, which

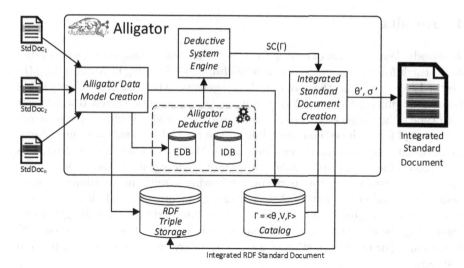

Fig. 1. The Alligator Architecture. ALLIGATOR receives documents in different Industry 4.0. standards, and creates an integrated document. Input documents are represented as RDF graphs and Datalog predicates (EDB) in the ALLIGATOR data model. Datalog intentional rules (IDB) characterize semantic heterogeneity types. A bottom-up evaluation of the Datalog program identifies semantic heterogeneity inconsistencies among input documents

is an XML-based standard, is translated into RDF using Krextor [11], an XSLT-based framework for converting XML to RDF. In addition, the mapping rules for the conversion using Krextor have to be created according to the AutomationML vocabulary, and an RDFS vocabulary[1] and the mapping rules[2] for the AutomationML standard have been defined. Similarly, we plan to define more rules to cover other standards and to semantically integrate them.

We present an example of one of the designed Datalog rules that states when two AutomationML elements are the same. Based on the AutomationML specification, this condition is met depending on links with external standards such as eCl@ss [3], which semantically identify elements. Accordingly, if two elements contains the same `eClassIRDI` as a value, then they are semantically equivalent. Listing 1.1 is a Datalog rule that represents this knowledge.

Listing 1.1. Rule 1.1: Semantic equivalence of two eClassIRDI AML attributes

```
sameEClassIRDI(X,Y) :- hasAttributeName(X, 'eClassIRDI') &
                       hasAttributeName(Y, 'eClassIRDI') &
                       hasAttributeValue(X,Z) &
                       hasAttributeValue(Y,Z).
```

---

[1] https://w3id.org/i40/aml/.

[2] https://raw.githubusercontent.com/EIS-Bonn/krextor/master/src/xslt/extract/aml.xsl.

4   Results

Testbeds. With the aim of testing the effectiveness of ALLIGATOR, we generated 30 Testbeds for *intra-standard interoperability* issues focusing in AutomationML. Testbeds are based on the semantic heterogeneity types M2 (granularity), M3 (schematic differences), and M6 (grouping and aggregation); ten testbeds per each type of heterogeneity. First, a seed (AutomationML document) was manually created for each testbed according to the type of semantic mapping. Next, we automatically generated two AutomationML documents derived from this seed containing a random number of semantic equivalent AutomationML elements[3]. The generation was performed following a uniform distribution. Testbeds corresponded to pairs of AutomationML documents, and thirty testbeds were evaluated in the study[4]. Further, a Gold Standard was manually generated computing the elements that were semantically equivalent as well as those different ones. For the compilation of a Gold Standard, we relied on the generated testbeds.

Table 1 reports on the values of these metrics for each type of semantic heterogeneity, i.e., M2, M3, and M6. We observed that for these semantic heterogeneity types, the value for precision is 1.0, i.e., ALLIGATOR correctly detected all the

Table 1. Effectiveness of Alligator. Per semantic heterogeneity type, effectiveness of ALLIGATOR is reported. In all testbeds, precision is 1.0. ALLIGATOR exhibits the highest performance in the testbeds of type M2 (F-measure is always 1.0), while in M3 and M6, the F-measure values are at least 0.8.

Granularity (M2)	TB1	TB2	TB3	TB4	TB5	TB6	TB7	TB8	TB9	TB10
Precision	1.0	1.0	1.0	1.0	1.0	1.0	1.0	1.0	1.0	1.0
Recall	1.0	1.0	1.0	1.0	1.0	1.0	1.0	1.0	1.0	1.0
F-measure	1.0	1.0	1.0	1.0	1.0	1.0	1.0	1.0	1.0	1.0

Schematic (M3)	TB1	TB2	TB3	TB4	TB5	TB6	TB7	TB8	TB9	TB10
Precision	1.0	1.0	1.0	1.0	1.0	1.0	1.0	1.0	1.0	1.0
Recall	1.0	1.0	1.0	1.0	1.0	1.0	0.83	1.0	0.88	0.75
F-measure	1.0	1.0	1.0	1.0	1.0	1.0	0.90	1.0	0.94	0.85

Grouping (M6)	TB1	TB2	TB3	TB4	TB5	TB6	TB7	TB8	TB9	TB10
Precision	1.0	1.0	1.0	1.0	1.0	1.0	1.0	1.0	1.0	1.0
Recall	1.0	1.0	1.0	0.66	1.0	1.0	1.0	1.0	1.0	0.83
F-measure	1.0	1.0	1.0	0.80	1.0	1.0	1.0	1.0	1.0	0.90

---

[3] https://github.com/i40-Tools/AMLGoldStandardGenerator.
[4] https://github.com/i40-Tools/HeterogeneityExampleData.

semantically equivalent elements. Further, recall and F-measure are also 1.0 in the testbeds of semantic heterogeneity M2. These results suggest that ALLIGA-TOR rules capture the knowledge required to *accurately* solve the *AutomationML semantic equivalent elements Identification* problem. For the semantic heterogeneity types M3 and M6, ALLIGATOR rules are not completely covering all possible semantic equivalences generated between *nested structures* of AutomationML elements. Thus, ALLIGATOR could not identify at most two semantic equivalent elements in five out of 20 testbeds of type M3 and M6.

5 Methodology

The methodology adopted in this doctoral work comprises the following tasks:

1. Investigation of state-of-the-art approaches relevant to the problem of integrating Industry 4.0 standards.
2. Formalization of the problem of integrating Industry 4.0 standards and proposed solutions; definition of research questions and hypotheses of our formal and empirical study are stated. ALLIGATOR is designed under the following *hypothesis*: an approach combining Datalog rules and Semantic Web technologies, for the integration of Industry 4.0 standards will exhibit better performance than the state of the art approaches. In addition, we identified the following research questions: (**RQ1**) Is ALLIGATOR able to identify pairs of semantic equivalent elements in Industry 4.0 documents? (**RQ2**) Does ALLIGATOR exhibit equal behavior whenever different types of semantic heterogeneity occur during the integration of Industry 4.0 documents?
3. Empirical evaluation of our hypothesis to measure ALLIGATOR performance with respect to state-of-the-art approaches.
   (a) Implementation of state-of-the-art or baselines approaches.
   (b) Definition of benchmarks to evaluate the proposed solutions.
   (c) Design and execution of experiments and statistical tests to validate or falsify our hypotheses. The result section was a first attempt to measure the effectiveness of ALLIGATOR. We designed a controlled experiment where some of the heterogeneity types were measured. Based on this idea we plan to extend the Testbeds to cover the following:
      i. All the heterogeneity types, i.e., from M1 to M7 for the *intra-* as well as for the *inter*-standard interoperability issues.
      ii. A validation with domain experts to evaluate results of the integrated documents, as well as the generated explanations. This validation will be conducted for both problems, *intra-* and *inter-* standard interoperability issues, and for all the heterogeneity types.
   (d) Analysis and discussion of the observed results.

6 Conclusions and Future Work

The problem of assessing *intra-* and *inter-standard* interoperability is addressed in this work, and we propose the combination of Deductive databases and Semantic Web technologies to effectively solve these problems. As a proof-of-concept, we present ALLIGATOR, a deductive framework for the integration of Industry 4.0 Standard documents. ALLIGATOR relies on Datalog and RDF to *accurately* represent the knowledge characterizing different types of semantic heterogeneity for documents described in Industry 4.0 standards. Currently, the main focus of our approach is to solve semantic heterogeneity issues that may occur among documents defined in the same, i.e., *intra-standard interoperability* problem. Nevertheless, we additionally plan to extend ALLIGATOR to integrate different standards and assess the *inter-standard interoperability* problem.

The results of the empirical evaluation for the *intra-standard interoperability* problem using AutomationML standard, suggest that ALLIGATOR is able to effectively solve the problems of *AutomationML semantic equivalent Identification element* and exhibits similar behavior for the three studied semantic heterogeneity types, i.e., granularity (M2), schematic (M3), and grouping (M6). In the future, we will empower the ALLIGATOR Deductive System Engine with the expressiveness of Datalog with negation and built-in predicates. Thus, ALLIGATOR will be able to represent all the other types of semantic heterogeneity in AutomationML, i.e., from M1 to M7. The heterogeneity types will be also defined for the *inter-interoperability* problem, i.e., between AutomationML and OPC UA. Further, we will extend ALLIGATOR to create explanations of the aligned elements of the integrated Industry 4.0 standard documents. Finally, we envision to develop a more general framework, capable of semantically integrate documents expressed in different Industry 4.0 standards.

Acknowledgments. The author would like to thank to Sören Auer and Maria-Esther Vidal for their guidance and fruitful discussions during the development of this doctoral work.

References

1. Biffl, S., Kovalenko, O., Lüder, A., Schmidt, N., Rosendahl, R.: Semantic mapping support in AutomationML. In: ETFA, pp. 1–4. IEEE (2014)
2. Ceri, S., Gottlob, G., Tanca, L.: What you always wanted to know about datalog (and never dared to ask). IEEE Trans. Knowl. Data Eng. **1**(1), 146–166 (1989)
3. eClass e.V.: eCl@ss standardized material and service classification (2016)
4. Enste, U., Mahnke, W.: OPC unified architecture. Automatisierungstechnik **59**(7), 397–405 (2011)
5. e.V. AutomationML, OPC Foundation: OPC UA information model for AutomationML. Status report (2016)
6. Grangel-González, I., Collarana, D., Halilaj, L., Lohmann, S., Lange, C., Vidal, M.-E., Auer, S.: Alligator: a deductive approach for the integration of Industry 4.0 standards. In: Blomqvist, E., Ciancarini, P., Poggi, F., Vitali, F. (eds.) EKAW 2016. LNCS (LNAI), vol. 10024, pp. 272–287. Springer, Cham (2016). doi:10.1007/978-3-319-49004-5_18
7. Henßen, R., Schleipen, M.: Interoperability between OPC UA and AutomationML. In: Procedia CIRP 8th International Conference on Digital Enterprise Technology DET, vol. 25 (2014)

8. Himmler, F.: Function based engineering with automationml - towards better standardization and seamless process integration in plant engineering. In: 12th International Conference on Tagung Wirtschaftsinformatik, WI (2015)
9. Kovalenko, O., Euzenat, J.: Semantic matching of engineering data structures. In: Biffl, S., Sabou, M. (eds.) Semantic Web for Intelligent Engineering Applications. Springer, Cham (2016)
10. Kovalenko, O., Wimmer, M., Sabou, M., Lüder, A., Ekaputra, F.J., Biffl, S.: Modeling AutomationML: semantic web technologies vs. model-driven engineering. In: 20th IEEE Conference on Emerging Technologies & Factory Automation, ETFA, pp. 1–4 (2015)
11. Lange, C.: Krextor - an extensible XML→RDF extraction framework. In: Scripting and Development for the Semantic Web, SFSW, vol. 449. CEUR Workshop Proceedings, Aachen, May 2009
12. Panetto, H., Zdravkovic, M., Jardim-Gonçalves, R., Romero, D., Cecil, J., Mezgár, I.: New perspectives for the future interoperable enterprise systems. Comput. Ind. **79**, 47–63 (2016)
13. Persson, J., Gallois, A., Björkelund, A., Hafdell, L., Haage, M., Malec, J., Nilsson, K., Nugues, P.: A knowledge integration framework for robotics. In: 41st International Symposium on Robotics and ROBOTIK 2010 (2010)
14. Sabou, M., Ekaputra, F., Kovalenko, O., Biffl, S.: Supporting the engineering of cyber-physical production systems with the AutomationML analyzer. In: 1st International Workshop on Cyber-Physical Production Systems, CPPS, pp. 1–8. IEEE (2016)
15. Schleipen, M., Damm, M., Henßen, R., Lüder, A., Schmidt, N., Sauer, O., Hoppe, S.: OPC UA and AutomationML-collaboration partners for one common goal: Industry 4.0. (2014)
16. Schleipen, M., Gutting, D., Sauerwein, F.: Domain dependant matching of MES knowledge and domain independent mapping of AutomationML models. In: 2012 IEEE 17th Conference on Emerging Technologies & Factory Automation, ETFA, pp. 1–7. IEEE (2012)
17. Schleipen, M., Okon, M.: The CAEX tool suite - user assistance for the use of standardized plant engineering data exchange. In: 15th IEEE International Conference on Emerging Technologies and Factory Automation, ETFA (2010)
18. Schmidt, N., Lüder, A., Rosendahl, R., Ryashentseva, D., Foehr, M., Vollmar, J.: Surveying integration approaches for relevance in cyber physical production systems. In: ETFA, pp. 1–8. IEEE (2015)

Extracting Knowledge Claims for Automatic Evidence Synthesis Using Semantic Technology

Jinlong Guo[✉]

School of Information Sciences, University of Illinois at Urbana-Champaign, Champaign, USA
jguo24@illinois.edu

Abstract. Systematic review, a form of evidence synthesis that critically appraises existing studies on the same topic and synthesizes study results, helps reduce the evidence gap. However, keeping the systematic review up-to-date is a great challenge partly due to the difficulty in interpreting the conclusion of a systematic review. A promising approach to this challenge is to make semantic representation of the claims made in both the systematic review and the included studies it synthesizes so that it's possible to automatically predict whether the conclusion of a systematic review changes given a new study. In this dissertation work, we developed a taxonomy to represent knowledge claims both in systematic review and its included studies with the goal of automatically updating a systematic review. We then developed machine learning models to automatically predict a synthesized claim from claims in individual studies.

Keywords: Systematic review · Evidence synthesis · Knowledge representation · Knowledge claim

1 Problem Statement

With the rapid increase in medical literature, it is increasingly difficult to keep up with the state-of-the-art medical evidence [1]. Systematic review is a type of literature review that collects and critically analyzes multiple research studies or papers, thus is able to resolve contradictory findings in the literature which is not uncommon [2]. Systematic reviews of randomized controlled trials are key in the practice of evidence-based medicine, which is regarded as the highest-form of medical evidence [3]. However, creating a systematic review is very time-consuming (may take months to years). Consequently, only a minority of trials have been assessed in systematic reviews [1]. Using current methods, the Cochrane Collaboration (the largest group in the world that produces systematic reviews) has not been able to keep even half of its reviews up-to-date [4], and other organizations are in a similar predicament [5]. Therefore, we need leaner and more efficient methods of staying up-to-date with the evidence.

Currently semi-automatic methods have been studied to facilitate the process of manual systematic review creation, from article screening [6, 7], data extraction [8] to quality [9] and risk of bias [10] rating. However, the biggest challenge of keeping systematic review (and other synthesized evidence like clinical practice guideline) up-to-date is an unsolved problem. The biggest challenge for updating a systematic review

© Springer International Publishing AG 2017
P. Ciancarini et al. (Eds.): EKAW 2016 Satellite Events, LNAI 10180, pp. 238–244, 2017.
DOI: 10.1007/978-3-319-58694-6_37

is to judge whether the update will influence the review findings or credibility sufficiently to justify the effort in updating it. This requires two critical components: a clear understanding of what is being claimed in the systematic review and what is being claimed in the new literature. Both components are extremely challenging as correctly interpreting neither the conclusion of systematic review [11] nor original experimental study [12] is a trivial task.

To deal with this challenge, living systematic reviews have been proposed, where high quality, online summaries of health research are updated as new research becomes available [13]. To realize such a living systematic review, semantic technology that supports a fine-granular representation of claims is required to detect a change in the conclusion [14]. The key technology that enables such a vision has not been fully realized. Our goal in this project is to develop automated techniques that will realize the vision of a living systematic review. Specifically, we will (a) develop a scheme to represent claims of conclusion for both a systematic review and its included (cited) studies, with a goal to automatically update or prioritize systematic reviews; (b) automate the representation scheme by developing models to predict synthesized claims from individual studies based on the claim representation.

2   State of the Art

The majority of work that attempts to automate systematic reviews focuses on identifying relevant studies [6, 15] rather than focusing on the conclusion of a study. Current approaches for updating a systematic review rely heavily on expert involvement [16]. While semantic technologies [14] have been suggested to enable a living systematic review, little work has been done to realize such a vision. The Cochrane group has initiated a linked data project[1], aiming to provide semantic support in accessing systematic reviews. However, so far they can only provide population, intervention, comparison and outcome (PICO) level annotation for the review [17], which is not able to balance the results from multiple studies and detect where a conclusion would change. In terms of claim representation, quite a lot of models [18–20] have been developed based on scientific literature, aiming to facilitate knowledge retrieval and discovery. However, these claim representations have not been applied on systematic reviews to facilitate interpreting and updating systematic reviews.

3   Proposed Approach

3.1   Overall Approach

Our approach is to extend the claim representation [19] that was developed to identify the outcomes of an individual empirical study to capture the conclusions of a systematic review and the relationship between the systematic review and the included studies. By representing the claims in a structured way, our model can automatically predict synthesized

---

[1] http://linkeddata.cochrane.org/.

claims (e.g. in a systematic review) from a group of relevant studies, which has the potential to detect signals that could change the conclusion of a systematic review.

Figure 1 shows the overall approach of our automatic evidence synthesis model using the extended claim representation. We will use the claim representation to annotate claims in both a systematic review and its included articles. We will then develop a machine learning model to automatically predict synthesized claim from claims in individual studies. Note that the "high", "moderate" and "low" labels in the synthesized claims represent the strength of the synthesized claim. The claim representation we developed could serve at least two purposes: First, only by representing the systematic review conclusions in a structured way, is it possible to detect the need to update the systematic review when new evidence appears. Second, representing the conclusion of a systematic review in a formalized way has the potential to reduce the gap in interpreting the findings of a systematic review.

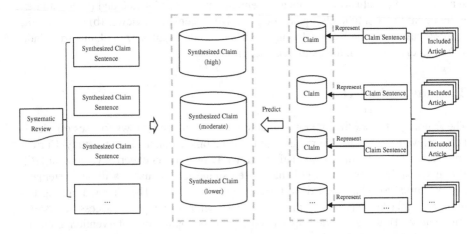

Fig. 1. Overall approach of automatic evidence synthesis using claim representation

3.2 Methodology

To implement the proposed approach above, several steps are required:

(1) **Developing the claim representation**: Based on the Claim Framework [19], a pilot study will be conducted to discover the claim types and other aspects of the claim such as direction of effect and strength of evidence. A large scale annotation study will then finalize the claim taxonomy developed.

(2) **Annotating the claim**: Extract synthesized claim sentences from systematic reviews and annotate them based on the claim representation developed. Extract the corresponding claim sentences from primary studies and annotate them based on the claim representation.

(3) **Developing models to predict synthesized claims in systematic reviews from claims in individual studies**: Study what factors might impact the conversion from primary claims to the synthesized claims so that we can use these features to

automatically predict the synthesized claims. Possible factors include (a) the direction and strength of claims; (b) the type of claims; (c) the semantic relationship between entities in the claims; (d) the location of claims in primary study (e.g. abstract or full text).

(4) **Automating the model**: Based on features discovered from previous steps, we will develop machine learning models to automatically predict synthesized claims from primary claims. Stanford Parser will be used to generate compound noun phrases as candidate entities in the claim. The UMLS ontology will be used for reconciling entities and relations in the claim. Textual entailment techniques will be considered for this task.

(5) **Evaluating the model**: To evaluate the synthesized claim prediction model, we will leave some annotated systematic reviews as test data and report the precision, recall and F value of the system compared with baseline (e.g. bag of words model). We will also use a situated case study to evaluate whether our model is able to detect the need for updating a systematic review.

4   Results and Discussion

So far we have conducted a case study on a collection of 79 systematic reviews from the Breast Cancer Cochrane Collaboration Group, which contain 2858 included studies. We have developed a representation for both synthesized and primary claims. Figure 2 shows the basic structure of the developed claim taxonomy. The claim types are based on the Claim Framework [19]. Other aspects of the claim include direction of effect and strength of evidence, the combination of which constitutes basic conclusion type of a systematic review [11]. Table 1 gives some specific examples from the reviews in our dataset to show the claim sentences with different strength of evidence in the context of evidence-based medicine.

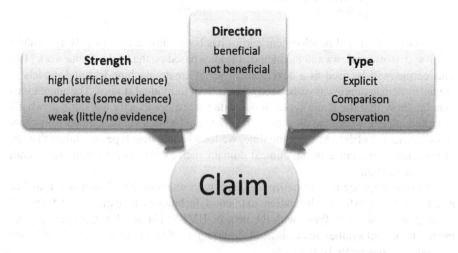

Fig. 2. Claim taxonomy for automatic evidence synthesis

Table 1. Example of claim sentences with different strength

Claim strength	Example	Explanation	Review PMID
High (sufficient evidence)	"Using CSFs significantly reduced the proportion of patients with FN (RR 0.27; 95% CI 0.11 to 0.70; number needed to treat for an additional beneficial outcome (NNTB) 12) but there was substantial heterogeneity which can be explained by possible differential effects of G-CSFs and GM-CSFs and different definitions of FN"	This level of evidence is towards definitive, though there's heterogeneity, but could be explained	23076939
Moderate (some evidence)	"There is evidence, though less reliable, of a decrease of all-cause mortality during chemotherapy and a reduced need for hospital care"	Some evidence means even though the pooled outcome reached statistical significance, but the events are mainly from on study, therefore, it's possible that future study could alter the conclusion	23076939
Weak (little/no evidence)	"No reliable evidence was found for a reduction of infection-related mortality, a higher dose intensity of chemotherapy with CSFs or diminished rates of severe neutropenia and infections"	Little or no evidence usually means the pooled estimate did not reach statistical significance or sometimes because of the low quality of included studies.	23076939

A manually created development corpus has been constructed comprising a subset of the systematic reviews and the included studies based on the Claim Framework [19]. This corpus will be used as a development set to create machine learning models to predict systematic review (synthesized) claims from individual studies. The annotations thus far show that the claim types in our date set are not exactly the same as what's proposed in [19], where the most common types are Explicit claim, Comparison claim and Observation claim. At the same time, we found some new types of claims that are of particular importance in the clinical domain such as side effect claim and overall effectiveness claim.

Our next steps are to (1) finalize the claim representation based on a new round of annotations; (2) extend information extraction techniques to extract key facets of existing and new claim types with the help of UMLS; (3) develop machine learning models to predict synthesized claims; (4) test our model in a new data set; (5) make the annotated corpus publicly available.

5   Conclusions and Future Work

Keeping up with the state of evidence is a great challenge. In this dissertation work, we propose a claim-based approach to address this challenge by automatically predicting synthesized conclusions (e.g. systematic review) from primary studies. The underlying key technology is the semantic representation of claims. Based on existing work about claim representation, we developed a claim taxonomy for the purpose of automatic evidence synthesis. Current annotation study shows the feasibility of this representation and automatic models are under development.

While text mining and natural language processing techniques have been widely used to facilitate the manual process of systematic review/evidence synthesis, we believe that our study is the first of its kind to examine the possibility of automatically updating/ prioritizing the systematic review at the conclusion level. While a systematic review is much more than just extracting data from the articles (e.g. critical appraisal is needed to rate the quality of the article), we took the first step towards automating the time consuming systematic review creation process by using a fine-granular representation of claims. In the future, if we can combine the claim extraction, automatic study quality prediction [9] and risk of bias prediction [10] together, we are closer to realizing full automation of systematic reviews, which will greatly accelerate evidence-based medicine and realize the vision of living systematic reviews [13].

Acknowledgement. The author would like to thank Professor Catherine Blake and Professor Jodi Schneider for their guidance and support in writing this paper.

References

1. Bastian, H., Glasziou, P., Chalmers, I.: Seventy-five trials and eleven systematic reviews a day: how will we ever keep up? PLoS Med. **7**, e1000326 (2010)
2. Ioannidis, J.P.: Why most published research findings are false. PLoS Med. **2**, e124 (2005)
3. Burns, P.B., Rohrich, R.J., Chung, K.C.: The levels of evidence and their role in evidence-based medicine. Plast. Reconstr. Surg. **128**, 305–310 (2011)
4. Koch, G.: No improvement–still less than half of the Cochrane reviews are up to date. In: XIV Cochrane Colloquium, Dublin (2006)
5. Garritty, C., Tsertsvadze, A., Tricco, A.C., Sampson, M., Moher, D.: Updating systematic reviews: an international survey. PLoS ONE **5**, e9914 (2010)
6. Cohen, A.M., Hersh, W.R., Peterson, K., Yen, P.-Y.: Reducing workload in systematic review preparation using automated citation classification. J. Am. Med. Inform. Assoc. **13**, 206–219 (2006)
7. Cohen, A.M., Smalheiser, N.R., McDonagh, M.S., Yu, C., Adams, C.E., Davis, J.M., Yu, P.S.: Automated confidence ranked classification of randomized controlled trial articles: an aid to evidence-based medicine. J. Am. Med. Inform. Assoc. **22**, 707–717 (2015)
8. Blake, C., Lucic, A.: Automatic endpoint detection to support the systematic review process. J. Biomed. Inform. **56**, 42–56 (2015)
9. Kilicoglu, H., Demner-Fushman, D., Rindflesch, T.C., Wilczynski, N.L., Haynes, R.B.: Towards automatic recognition of scientifically rigorous clinical research evidence. J. Am. Med. Inform. Assoc. **16**, 25–31 (2009)

10. Marshall, I.J., Kuiper, J., Wallace, B.C.: RobotReviewer: evaluation of a system for automatically assessing bias in clinical trials. J. Am. Med. Inform. Assoc. **23**, 193–201 (2016)

11. Lai, N.M., Teng, C.L., Lee, M.L.: Interpreting systematic reviews: are we ready to make our own conclusions? A Cross Sect. Study BMC Med. **9**, 30 (2011)

12. Boutron, I., Dutton, S., Ravaud, P., Altman, D.G.: Reporting and interpretation of randomized controlled trials with statistically nonsignificant results for primary outcomes. JAMA **303**, 2058–2064 (2010)

13. Elliott, J.H., Turner, T., Clavisi, O., Thomas, J., Higgins, J.P., Mavergames, C., Gruen, R.L.: Living systematic reviews: an emerging opportunity to narrow the evidence-practice gap. PLoS Med. **11**, e1001603 (2014)

14. Slaughter, L., Berntsen, C.F., Brandt, L., Mavergames, C.: Enabling living systematic reviews and clinical guidelines through semantic technologies. D-Lib Mag. **21**, 8 (2015)

15. Cohen, A.M., Ambert, K., McDonagh, M.: Cross-topic learning for work prioritization in systematic review creation and update. J. Am. Med. Inform. Assoc. **16**, 690–704 (2009)

16. Shekelle, P.G., Motala, A., Johnsen, B., Newberry, S.J.: Assessment of a method to detect signals for updating systematic reviews. Syst. Rev. **3**, 13 (2014)

17. Mavergames, C., Oliver, S., Becker, L.: Systematic reviews as an interface to the web of (trial) data: using PICO as an ontology for knowledge synthesis in evidence-based healthcare research (2013)

18. de Waard, A., Buckingham Shum, S., Carusi, A., Park, J., Samwald, M., Sándor, Á.: Hypotheses, evidence and relationships: the HypER approach for representing scientific knowledge claims (2009)

19. Blake, C.: Beyond genes, proteins, and abstracts: identifying scientific claims from full-text biomedical articles. J. Biomed. Inform. **43**, 173–189 (2010)

20. Clark, T., Ciccarese, P.N., Goble, C.A.: Micropublications: a semantic model for claims, evidence, arguments and annotations in biomedical communications. J. Biomed. Semant. **5**, 1 (2014)

A Proposal for Self-Service OLAP Endpoints for Linked RDF Datasets

Median Hilal[(✉)] [iD]

Department of Business Informatics – Data & Knowledge Engineering,
Johannes Kepler University, Linz, Austria
hilal@dke.uni-linz.ac.at

Abstract. Leveraging external RDF data for OLAP analysis opens a wide variety of possibilities that enable analysts to gain interesting insights related to their businesses. While variations of statistical linked data are easily accessible to OLAP systems, exploiting non-statistical linked data, such as DBpedia, for OLAP analysis is not trivial. An OLAP system for these data should, on the one hand, take into account the big volume, heterogeneity, graph nature, and semantics of the RDF data. On the other hand, dealing with external RDF data requires a degree of self-sufficiency of the analyst, which can be met via self-service OLAP, without assistance of specialists. In this paper, we argue the need for self-service OLAP endpoints for linked RDF datasets. We review the related literature and sketch an approach. In particular, we propose the use of multidimensional schemas and analysis graphs over linked RDF datasets, which will empower users to perform self-service OLAP analysis on the linked RDF datasets.

Keywords: Resource Description Framework · Online Analytical Processing · Self-Service Business Intelligence

1 Introduction

Online Analytical Processing (OLAP) supports managers making business decisions. Organizing data in multidimensional (MD) schemas which consist of dimensions, facts, and measures, allows for the analysis of the data using OLAP tools. In order to obtain new perspectives on their businesses, organizations are increasingly including external (web) data into their local analysis. Web data from various domains are published using the Resource Description Framework (RDF) [13]. Open RDF datasets can be linked to each other, leading to Linked Open Data (LOD). OLAP analysis using LOD commonly exploits variations of statistical linked data which constitute only a portion of the available LOD. LOD represent a rich resource of knowledge. RDF data, however, present certain particularities with respect to the data traditionally used for OLAP, that is, these data are heterogeneous, dynamic, big-volume, semantically-rich, graph-structured, and don't follow MD structure. As a consequence, employing

© Springer International Publishing AG 2017
P. Ciancarini et al. (Eds.): EKAW 2016 Satellite Events, LNAI 10180, pp. 245–250, 2017.
DOI: 10.1007/978-3-319-58694-6_38

these data for OLAP analysis is not trivial and requires acknowledging the specific features of these data. Furthermore, performing analysis over external RDF datasets[1] requires users to have sufficient knowledge of both data semantics and querying mechanisms. Self-service business intelligence (SSBI) is thus necessary in the context of OLAP with RDF datasets. SSBI enables users that are non-experts in OLAP technology to perform data analysis without intervention of specialists while rendering expert users more productive [2]. Alpar and Schulz [2] distinguish three levels of self-service: "(i) usage of information, (ii) creation of information, and (iii) creation of information resources". Those levels correspond to increasing self-reliance of the user and system support.

As a motivating example, "Linked Data Integration for Global Publishers"[2] semantically annotates news from different websites using a dedicated ontology. The data are stored as RDF, linked to LOD concepts, and a SPARQL endpoint is provided. Consider *Anna* a journalist from a media company, who is interested in performing some investigations using these data. For example, *Anna* wants to know how the Tesla automobile crash[3] was reflected in the news by comparing count of the daily mentions involving Tesla in that specific time period. *Anna* has to manually access the SPARQL endpoint, discover entities and relationships, come up with suitable queries and pose them to the endpoint. This is a cumbersome and misleading process which is only possible if *Anna* is experienced in SPARQL, OLAP, and the dataset structure. In our vision, a third party, that is interested in providing such analyses, can design MD schema virtually over the dataset and provide dynamic self-service means with the schema. Thus, *Anna* can easily access a dedicated OLAP endpoint, select her interest within the MD schema, and instantiate the predefined analysis patterns to fit her needs. The operations required to fulfill this process are performed in the background.

The research question raised is *how to empower various users to perform self-service OLAP over RDF datasets in a way that, on one hand, recognizes the specific features of the RDF data and, on the other hand, supports users by means of self-service.* Correspondingly, we define a set of requirements:

R1. Adaption to big-volume, dynamic, unstable and non-controllable external RDF datasets, which may imply in-place processing [6].
R2. Employing the semantics of the RDF data in the analysis process, especially subsumption hierarchies which may lead to new insights over the data.
R3. Considering the graph nature of the RDF data in the analysis process.
R4. Coping with the heterogeneity of the RDF data.
R5. Supporting variation of subjects and contexts of the analytical possibilities.

---

[1] Though our main motive is LOD, the principles presented apply to other variations of RDF datasets. Thus, we use the term RDF datasets in general from now on.
[2] http://ontotext.com/semantic-solutions/dynamic-semantic-publishing-platform/linked-data-integration-for-global-publishers/.
[3] https://www.theguardian.com/technology/2016/jun/30/tesla-autopilot-death-self-driving-car-elon-musk.

R6. Empowering users by means of self-service, especially at the first level. Users should be able to access/navigate prepared analyses without technical terms.

R7. Reuse, collaboration and sharing of the analyses among users [1].

R8. Enabling the explicit expression of MD hierarchical structures.

The remainder of this paper is organized as follows. Section 2 presents the state of the art. Section 3 sketches our proposed approach towards self-service OLAP endpoints for linked RDF datasets. Section 4 concludes the paper with a summary and an outlook on future work.

2   State of the Art

We categorize methods that perform OLAP-style analysis over RDF data in three main categories. The first category includes MD integrated ontologies [8] and the semantic data warehouse [7], which considered extracting semantic web data, loading them into a local data warehouse, designing the analysis, retrieving instances, and performing classical OLAP. This method is mainly directed toward local warehousing and relational OLAP and not suitable for dynamic and big volume data. Furthermore, no means of self-service are presented at the first level. The second category includes methods employing specific RDF(S) vocabularies for OLAP analysis where OLAP cubes can be published over the web, and OLAP operations can be implemented over them. The data cube vocabulary (QB) [12] enables publishing statistical linked data. To overcome the problem that QB is not originally dedicated for OLAP, the Open Cube vocabulary (OC) [5] was proposed to specify and publish MD cubes, so that OLAP operations are realized through SPARQL queries. However, to support reusing data previously published in QB format, which OC does not satisfy, QB4OLAP vocabulary [3] extended QB in order to support OLAP and reuse QB data at the same time. The QB4OLAP vocabulary was further extended [4] to support the representation of dimension hierarchies. QB2OLAP [11] presented a tool to perform OLAP over already published QB data and allow transforming QB into QB4OLAP. Although these methods use a model tailored to RDF data, they are only applicable to data originally published in a corresponding vocabulary. Additionally, only QB2OLAP [11] presents a sense of self-service to extend the multi-dimensional schemas. The third category includes the analytical schemas [10], which are graphs where each node and edge is a view over the original RDF data. These schemas can be analytically queried by means of rooted basic graph pattern and plans are proposed to answer such queries. Analytical schemas, however, lack the MD structure, which makes it harder for users to query these schemas and for specialists to provide guidance to users.

3   Proposed Approach

We propose to enable self-service OLAP analysis over RDF datasets via two main notions. First, *Multidimensional Schemas* (MDSs) facilitate expressing analytical possibilities and performing OLAP analysis over arbitrary RDF datasets.

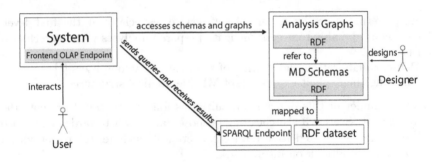

Fig. 1. Approach overview

Second, *Analysis Graphs* (AGs) enable self-service OLAP over these MDSs by expressing patterns of analytical processes that can be instantiated by users.

Multidimensional schemas are superimposed over the datasets and basically follow the MD model (requirement **R8** in Sect. 1). Those schemas will be mapped to the dataset via SPARQL queries declaring how the instances of each MD element are retrieved, with no need in principle for materialization. This allows the data to be up-to-date (**R1**), and hides the technical details from users. MDSs will enable easily expressing various analytical possibilities over the same dataset with less need for maintenance and storage (**R5**). MDSs will allow including subsumption hierarchies in the schema by introducing specialization constructs (**R2**). This also enables modeling heterogeneity, where specialized elements can have their own properties (**R4**). Additionally, MDSs should be capable of expressing graph features (**R3**). An OWL ontology will be developed to express and publish MDSs so that they can be accessed and reused (**R7**).

Analysis graphs, originally proposed for relational warehousing [9], enable expressing experts' knowledge of the analysis process as patterns that could be instantiated by users, in response for a specific business need, goal, or event. Analysis graphs are stated on the schema level, which may contain variables. Users can use AGs on the instance level by binding the variables to what fits their interests. Thus, AGs empower users who have limited domain or technical knowledge (**R6**). A node of AG is an analysis situation corresponding to a MD query. An edge of AG is a navigation step corresponding to a set of OLAP operations. To realize AGs in our context, first they will be adapted and extended to achieve compatibility with MDSs. Then, a dedicated RDFS vocabulary will be developed to express and publish these graphs (**R7**).

To conduct an analysis using an AG, user selects an analysis situation to start from and binds its variables, and can then navigate to another analysis situation. Users can also use analysis situations and navigation operators to freely query a MDS with no need to use a predefined analysis graph. A bound analysis situation corresponds to a query. The result of the query execution is an OLAP cube. The user interacts with the system endpoint while the system accesses the MDSs and AGs, and sends queries and receives results from the dataset endpoint. Figure 1 simplifies our intended solution architecture. To the best of our knowledge, our approach is the first to tackle self-service OLAP at the first level against superimposed MD structures that are built over arbitrary RDF datasets.

4 Summary and Future Work

In this paper, we proposed our approach to performing OLAP analysis over RDF datasets in self-service manner. This approach will enable publishing lightweight, yet expressive MD schemas where a wide variety of users could easily exploit the thus described RDF datasets by means of self-service BI through AGs. Future work includes developing the model of the MDSs as well as adapting and extending AGs. Additionally, we plan to define formally how the MD elements are mapped to the underlying datasets, and how SPARQL queries are constructed from analysis situations and navigations. Thus, we will have all the items required to realize the OLAP endpoints. Furthermore, it is important to provide tool support to the designers of the MDSs and the AGs.

Acknowledgment. I thank my supervisors, Dr. Christoph Schuetz and Dr. Michael Schrefl. I am funded by Erasmus Mundus - ASSUR program.

References

1. Abelló, A., Darmont, J., Etcheverry, L., Golfarelli, M., Mazón López, J.N., Naumann, F., Pedersen, T.B., Rizzi, S., Trujillo Mondéjar, J.C., Vassiliadis, P., et al.: Fusion cubes: towards self-service business intelligence (2013)
2. Alpar, P., Schulz, M.: Self-service business intelligence. In: BISE 2016, vol. 58, no. 2, pp. 151–155 (2016)
3. Etcheverry, L., Vaisman, A.: QB4OLAP: a new vocabulary for OLAP cubes on the semantic web. In: Proceedings of COLD (2012)
4. Etcheverry, L., Vaisman, A., Zimányi, E.: Modeling and querying data warehouses on the semantic web using QB4OLAP. In: Bellatreche, L., Mohania, M.K. (eds.) DaWaK 2014. LNCS, vol. 8646, pp. 45–56. Springer, Cham (2014). doi:10.1007/978-3-319-10160-6_5
5. Etcheverry, L., Vaisman, A.A.: Enhancing OLAP analysis with web cubes. In: Simperl, E., Cimiano, P., Polleres, A., Corcho, O., Presutti, V. (eds.) ESWC 2012. LNCS, vol. 7295, pp. 469–483. Springer, Heidelberg (2012). doi:10.1007/978-3-642-30284-8_38
6. Maali, F., Decker, S.: Towards an RDF analytics language: learning from successful experiences. In: COLD 2013, vol. 1034, pp. 136–145. CEUR-WS.org (2013)
7. Nebot, V., Berlanga, R.: Building data warehouses with semantic web data. Decis. Support Syst. **52**(4), 853–868 (2012)
8. Nebot, V., Berlanga, R., Pérez, J.M., Aramburu, M.J., Pedersen, T.B.: Multidimensional integrated ontologies: a framework for designing semantic data warehouses. In: Spaccapietra, S., Zimányi, E., Song, I.-Y. (eds.) Journal on Data Semantics XIII. LNCS, vol. 5530, pp. 1–36. Springer, Heidelberg (2009). doi:10.1007/978-3-642-03098-7_1

9. Neuböck, T., Schrefl, M.: Modelling knowledge about data analysis processes in manufacturing. IFAC-PapersOnLine **48**(3), 277–282 (2015)
10. Roatis, A.: Efficient querying and analytics of semantic web data. Ph.D. thesis, Paris 11 (2014)
11. Varga, J., Etcheverry, L., Vaisman, A.A., Romero, O., Pedersen, T.B., Thomsen, C.: QB2OLAP: enabling OLAP on statistical linked open data. In: ICDE 2016, pp. 1346–1349. IEEE (2016)
12. W3C: The RDF data cube vocabulary. https://www.w3.org/TR/2012/WD-vocab-data-cube-20120405/
13. W3C: The resource description framework. http://www.w3.org/RDF

Ontology Learning from Software Requirements Specification (SRS)

Muhammad Ismail[✉]

Department of Computer Science and Informatics, School of Engineering,
Jönköping University, Jönköping, Sweden
muhammad.ismail@ju.com
http://www.ju.se

Abstract. Learning ontologies from software requirements specifications with individuals and relations between individuals to represent detailed information, such as input, condition and expected result of a requirement, is a difficult task. System specification ontologies (SSOs) can be developed from software requirement specifications to represent requirements and can be used to automate some time-consuming activities in software development processes. However, manually developing SSOs to represent requirements and domain knowledge of a software system is a time-consuming and a challenging task. The focus of this PhD is how to create ontologies semi-automatically from SRS. We will develop a framework that can be a possible solution to create semi-automatically ontologies from SRS. The developed framework will mainly be evaluated by using the constructed ontologies in the software testing process and automating a part of it. i.e. test case generation.

Keywords: Ontology · Ontology learning · Software requirements specification · Software development process

1 Problem Description

The process of software development includes many activities, such as requirements analysis, requirements validation and software verification, that are time consuming if not automated. One promising way to automate such time-consuming activities is to use ontologies, because ontologies are machine processable models, support reasoning that can be used to assert consistency of models, and can be used in all phases of the software development process [1–3].

System specification ontology (SSO) is an ontology that captures domain knowledge and knowledge about the software system that is being developed. As software requirements specifications are used in all phases, an SSO can be created from software requirements specifications, which can be used to support different activities in the software development process. For example, such SSO represents domain knowledge in machine processable format that could help in test case generation to support software testing [4,5].

© Springer International Publishing AG 2017
P. Ciancarini et al. (Eds.): EKAW 2016 Satellite Events, LNAI 10180, pp. 251–255, 2017.
DOI: 10.1007/978-3-319-58694-6_39

Ontologies can be developed manually, semi-automatically or automatically. Manual development of ontologies is a labor-intensive, time-consuming and expensive [6,7]. Ontology and domain experts process all information manually and should be participating throughout the manual ontology development process, therefore expensive. It is beneficial to create ontologies semi-automatically or automatically to save time and resources.

Ontology learning is a term given for the semi-automatic or automatic construction of ontologies. Ontology learning is a process of extracting terms, identifying concepts and instances[1], relations, and axioms from data sources to create and maintain an ontology [8].

There are existing ontology learning systems, such as OntoGen [9], Text2Onto [10] and CRCTOL [11], which have implemented different methods and algorithms for ontology learning. However, they focus on extracting concepts and relations between concepts, but can not extract individual and relations between individuals from software requirements specifications.

Software requirements specifications have detailed information about requirements such as inputs, conditions, actions and expected results. In software development, this detailed information about the requirements is required to implement the functionality and automate the processes. This Detailed information about the requirements can be specified by creating individuals and defining relations between individuals in an ontology. Individuals and relations between individuals are important for reasoning, inference rules and processing in applications for practical use of ontologies [4]. Existing ontology learning methods and systems need to be improved for creating ontologies from software requirements specifications to represent requirements and domain knowledge of a software system.

To summarize, it seems that the semi-automatic creation of ontologies from software requirements specifications is a challenging problem. The focus of this PhD project is how to generate ontologies semi-automatically from software requirements specifications to use in software development processes. In practice, manual ontology development is not feasible for the software industry because it requires time, resources and expertise which will not save time and cost in the software development processes.

2   State of the Art

During the last decades, several systems and tools have been developed for ontology learning. These systems and tools work with different types of data and information resources. Several systems, such as [9–12] have been developed for ontology learning from text. The OntoGen system [9] is a semi-automatic and data-driven ontology editor for creating ontologies. The system uses machine learning and text mining algorithms, and suggests concepts and relations to the

---

[1] Individuals and instances are same in our context. In OWL language, individuals are known as instances as well.

users. The users of the OntoGen system have full control in the ontology construction process by accepting or rejecting the system suggestions. The users can manually adjust the properties of the suggestions for the ontology. The approach is highly interactive and the users need to know the process of ontology development and knowledge of the domain where developed ontologies supposed to use.

The Text2Onto [10] is an ontology learning framework which combines machine learning and natural language processing (NLP) approaches to learn an ontology. The idea is to get better knowledge extraction results by using several different approaches and then combining the results. The OntoExtractor [13] is an ontology learning framework for extracting seed concepts and relations from natural language text to overcome problems related to ontology refinement, axiom extraction and utilizing core concepts. The CRCTOL [11] is a domain ontology learning system that used statistical and lexico-syntactic methods to extract key concepts from a document collection. Most of existing systems and tools use statistical methods to extract terms and concepts from text and are dependent on size of text, which can affect the performance of the methods for extracting terms and concepts for creating an ontology [14]. Moreover, the existing methods extract only salient concept and can not extract specific technical concepts contained in technical documents [15]. In our case, software requirements specifications are the technical documents that will be used to learn ontologies. It is important to extract all concepts including individuals of concepts that can be used to represent information for the practical use of ontologies.

3   Aim and Objectives

The overall aim of the work is to create ontologies semi-automatically from the software requirements specifications. To reach this aim, a number of objectives have been specified:

O1. *To understand the manual process of ontology development for software requirements specifications.*
O2. *Evaluate existing ontology learning methods, to determine which methods can contribute to create ontology from SRS semi-automatically.*
O3. *Develop a framework for creating system specification ontologies from SRS documents.*
   – O3.1. Develop methods to extract terms from SRS, formation of ontology elements such as, concepts and relations, and populate ontologies.
   – O3.2. Develop a proof of concept of the framework to create ontologies from SRS.
O4. *Evaluate the constructed framework and the methods within framework.*

4 Methodology

The following steps will be performed to meet the objectives and reach the aim.

- We will analyze software requirements specifications for a software systems within a particular organization.
- We will develop an ontology manually from the requirements specifications to analyze the ontology development process to make it semi-automatic.
- A systematic literature review (SLR) will be done for ontology learning from SRS to keep updated about relevant research in the field. Another reason for SLR is to investigate existing methods of ontology learning can partially be used to learn ontologies from SRS.
- We will develop a framework and a set of methods within framework and implement to create system specification ontologies (SSOs) from SRS semi-automatically.
- We will do evaluation of the developed framework, such as by using the constructed ontologies in the software testing process and automating test case generation, evaluating through correctness and usability of constructed ontologies.

5 Preliminary Results and Future Work

The current research is part of the OSTAG[2] project and the thesis is in an early stage. The analysis of software requirements specifications and related documents for a communication software is in progress. An ontology has been developed manually from software requirements specifications for the particular communication software. An evaluation of manually developed ontology has been done by using it in the software testing process and by automating test case generation. By developing an ontology manually and evaluating it in the OSTAG project, we have got a clear idea about the type of domain knowledge and information required for an ontology to be useful for test automation. The ontology developed manually will be used for the evaluation of semi-automatically constructed ontologies. Further, a literature review for ontology learning from text has been conducted as software requirements are often written in natural language text.

The next step is to investigate existing methods of ontology learning that partially can be used for creating ontologies from SRS. Afterwards, the development of the framework and a set of methods within framework to perform different ontology-learning tasks, such as terms and concepts extraction, creation of instances and relations between instances, will be developed to learn ontologies from SRS. A proof of concepts will be developed of the framework. Finally, the evaluation of framework will be done through different ways. The evaluation will be done by comparing results of the framework and methods with the existing solutions. An indirect evaluation of the framework will be done by using constructed ontologies in software testing process and automating test case generation.

---

[2] http://ju.se/en/research/research-groups/computer-science-and-informatics/seman tic-technologies/research-projects/ontology-based-software-test-case-generation-ostag.html.

Acknowledgments. I would like to thank and gratefully acknowledge my supervisors, Dr. Vladimir Tarasov, Dr. He Tan from School of Engineering, Jönköping university and Dr. Birgitta Lindström from Skövde university for their support, motivation, valuable comments and guidance. This work is part of OSTAG project, funded from KK-Foundation by grant KKS-20140170 and will be carried at School of Engineering, Jönköping University, Jönköping.

References

1. Hesse, W.: Ontologies in the software engineering process. In: EAI, pp. 3–16 (2005)
2. Happel, H.-J., Seedorf, S.: Applications of ontologies in software engineering. In: Proceedings of Workshop on Semantic Web Enabled Software Engineering (SWESE) on the ISWC, pp. 5–9. Citeseer (2006)
3. Zedlitz, J., Luttenberger, N.: Transforming between UML conceptual models and owl 2 ontologies. In: Terra Cognita 2012 Workshop, vol. 6, p. 15 (2012)
4. Tarasov, V., Tan, H., Ismail, M., Adlemo, A., Johansson, M.: Application of inference rules to a software requirements ontology to generate software test cases. In: Dragoni, M., Poveda-Villalón, M., Jimenez-Ruiz, E. (eds.) OWLED 2016, ORE 2016. LNCS, vol. 10161, pp. 82–94. Springer, Cham (2017). doi:10.1007/978-3-319-54627-8_7
5. Souza, É.F., Falbo, R.A., Vijaykumar, N.L.: Ontologies in software testing: a systematic. In: VI Seminar on Ontology Research in Brazil, p. 71 (2013)
6. Yijian, W., Zhang, S., Zhao, W.: Towards learning domain ontology from legacy documents. In: Fourth International Conference on Digital Society, ICDS 2010, pp. 164–171. IEEE (2010)
7. Zhou, L.: Ontology learning: state of the art and open issues. Inf. Technol. Manage. **8**(3), 241–252 (2007)
8. Wong, W., Liu, W., Bennamoun, M.: Ontology learning from text: a look back and into the future. ACM Comput. Surv. (CSUR) **44**(4), 20 (2012)
9. Fortuna, B., Grobelnik, M., Mladenic, D.: OntoGen: semi-automatic ontology editor. In: Smith, M.J., Salvendy, G. (eds.) Human Interface 2007. LNCS, vol. 4558, pp. 309–318. Springer, Heidelberg (2007). doi:10.1007/978-3-540-73354-6_34
10. Cimiano, P., Völker, J.: Text2Onto. In: Montoyo, A., Muñoz, R., Métais, E. (eds.) NLDB 2005. LNCS, vol. 3513, pp. 227–238. Springer, Heidelberg (2005). doi:10.1007/11428817_21
11. Jiang, X., Tan, A.-H.: Crctol: a semantic-based domain ontology learning system. J. Am. Soc. Inf. Sci. Technol. **61**(1), 150–168 (2010)
12. Drymonas, E., Zervanou, K., Petrakis, E.G.M.: Unsupervised ontology acquisition from plain texts: the *OntoGain* system. In: Hopfe, C.J., Rezgui, Y., Métais, E., Preece, A., Li, H. (eds.) NLDB 2010. LNCS, vol. 6177, pp. 277–287. Springer, Heidelberg (2010). doi:10.1007/978-3-642-13881-2_29
13. Nie, X., Zhou, J.: A domain adaptive ontology learning framework. In: IEEE International Conference on Networking, Sensing and Control, ICNSC 2008, pp. 1726–1729. IEEE (2008)
14. Browarnik, A., Maimon, O.: Departing the ontology layer cake. In: Modern Computational Models of Semantic Discovery in Natural Language, p. 167 (2015)
15. Park, J., Cho, W., Rho, S.: Evaluating ontology extraction tools using a comprehensive evaluation framework. Data Knowl. Eng. **69**(10), 1043–1061 (2010)

Managing Ontology Mapping Change Based on Changing Inference Sets

Matthias Jurisch[✉]

Computer Science Department,
RheinMain University of Applied Sciences,
Unter den Eichen 5, 65195 Wiesbaden, Germany
matthias.jurisch@hs-rm.de

Abstract. Dealing with ontology changes is of high importance for ontology engineers in different application domains. In some applications domain ontologies cover overlapping topics. Ontology mappings represent this overlap and allow specifying a connection between domain ontologies. Adapting the mappings when the domain ontologies are changed is a problem still lacking a thoroughly evaluated, systematic approach. This doctoral research aims at providing a new approach to this problem by comparing the inferences computed from the relevant regions of the domain ontologies and the mappings before and after changes occur. A change model can be used to categorize differences in the inferences as intentional or unintentional and propose solutions to adapt the mappings.

Keywords: Ontology mapping · Mapping management · Mapping adaption

1 Introduction

One of the most important aspects of the semantic web is the interlinking of models from different sources. This makes it easy for ontology engineers to reference information from several domains. When dealing with models that cover overlapping topics, the connection between the ontologies can be formulated using *ontology mappings* (sometimes called ontology alignments). An ontology mapping is a set of correspondences between entities from connected ontologies.

For finding these connections between ontologies, several approaches exist. Mappings can be discovered manually or automatically. An overview of approaches for finding these mappings automatically ("ontology matching") has been presented by Euzenat and Shvaiko [6,17]. It is currently unclear how to deal with changes in connected ontologies [17]. When ontologies change, mappings need to be kept up-to-date. Automated support for this task is essential to ontologies that have many mappings to other ontologies. This is the problem that will be approached in my doctoral research.

In the field of software engineering, ontologies can be used to integrate knowledge from different domains and phases of the development process to support

© Springer International Publishing AG 2017
P. Ciancarini et al. (Eds.): EKAW 2016 Satellite Events, LNAI 10180, pp. 256–263, 2017.
DOI: 10.1007/978-3-319-58694-6_40

a better communication between different departments. Possible domains where ontologies can be used to support software development projects are ontologies for organizational aspects [7], UX design artifacts [15], software architecture specifications [4] and possibly other domains. Ontology mappings can be used to support the information integration of software development data by describing the relation between ontology elements from different domains. This could improve the dissemination of semantic web technologies in this area. The author of this publication is currently participating in a research and development project with an industry partner that evaluates using mapped ontologies from the mentioned domains as a foundation for a recommender system that supports the software development process. These ontologies can change due to manually added information or ontology mining. Since recreating the mappings can be very time-consuming, adaption of the mappings to these changes is a very important task that could benefit from a semi-automatic approach. A new approach for the mapping adaption problem could benefit many application areas that have to deal with changing mapped ontologies (e.g. ontologies in life-sciences, ontology incoherency and ontologies in interwoven systems), but this work will focus on the described application domain in software engineering.

The remainder of this paper is structured as follows: A problem description is given in Sect. 2. Section 3 summarizes the state of the art and shows the research gap that is addressed by this doctoral work. Section 4 presents the proposed approach for the identified problem. The methodology to implement and evaluate the approach is given in Sect. 5. The current state of the work is described in Sect. 6 and a conclusion and outlook are given in Sect. 7.

2   Problem

The problem of adapting mappings between changing schemas is called the mapping adaption problem [19,20]. The doctoral research described in this paper addresses the mapping adaption problem for ontology mappings. Formally, we define a mapping between ontologies analogous to [8]: A mapping between two ontologies $\mathcal{O}_1$ and $\mathcal{O}_2$ is defined as

$$M_{\mathcal{O}_1,\mathcal{O}_2} = \{(c_1, c_2, semType) | c_1 \in \mathcal{O}_1, c_2 \in \mathcal{O}_2, semType \in \{\equiv, \leq, \geq\}\}$$

Each entry in the set $M_{\mathcal{O}_1,\mathcal{O}_2}$ is called mapping entry. Two versions of the same ontology changed over time are denoted by $\mathcal{O}$ and $\mathcal{O}'$. The mapping adaption problem for two ontologies $\mathcal{O}_1$ and $\mathcal{O}_2$ connected by $M_{\mathcal{O}_1,\mathcal{O}_2}$ can then be restated as finding a new mapping $M'_{\mathcal{O}_1,\mathcal{O}'_2}$, when $\mathcal{O}_2$ evolves to $\mathcal{O}'_2$ and

- $M_{\mathcal{O}_1,\mathcal{O}_2} \cap M'_{\mathcal{O}_1,\mathcal{O}'_2}$ is as large as possible and
- $M'_{\mathcal{O}_1,\mathcal{O}'_2}$ fulfills application specific properties (e.g. mappings may not cause inconsistencies, some classes may not be removed, mappings have to fulfill domain specific test cases).

Currently, this problem is usually solved manually by ontology engineers at the time when ontologies evolve. While some kind of user activity can not be omitted, performing this task completely manually is highly error-prone. Important mapping entries could be forgotten or obsolete ones kept inside the mapping, especially when the evolution is executed by inexperienced ontology engineers. This is caused by a lack of tool-support for finding inconsistencies and incoherences. Besides overlooking these issues, ontology engineers can not easily find out, which mapping entries are affected by changes without causing inconsistencies or incoherences. Especially in big ontologies, it can be difficult to manually determine which parts of an ontology mapping are still correct and which parts have to be adapted.

In my doctoral research, I want to address this problem by creating a systematic approach that helps keeping these mappings up-to-date when ontologies evolve. Ontology engineers shall be supported in finding and verifying new mappings based on old ones. This process shall incorporate as much automation as possible during the adaption of the old mapping.

3   State of the Art

The mapping adaption problem is not limited to ontology mappings. It has been discussed in database research as schema mapping adaption, where the discussion focussed on the adaption of mappings between XML- or database-schemas. Velegrakis et al. [19] proposed an *incremental approach* reacting to specific changes in the schemas based on rules. For each change pattern a specific modification for the mapping is defined. Yu et al. [20] proposed an approach that is based on a composition of mappings. A new mapping $M'_{\mathcal{O}_1,\mathcal{O}'_2}$ is created by a *composition of the mapping* $M_{\mathcal{O}_1,\mathcal{O}_2}$ and $M^+_{\mathcal{O}_2,\mathcal{O}'_2}$, the mapping between $\mathcal{O}_2$ and $\mathcal{O}'_2$. These approaches do not exploit the underlying semantics of the schemas.

Changes in ontologies in general have been studied in the field of ontology evolution, where different approaches have been proposed [21]. Currently, there are several methods for managing changes in single ontologies and several tools, that implement these in practice. From a description logics standpoint, there have also been approaches to formalize ontology evolution. While traditional belief change theory like AGM theory [1] can not be applied to description logics used in modern ontologies, a revised set of AGM constraints that could be used to allow coherent revision of ontologies has been proposed [16]. These approaches might allow a consistent evolution of ontologies, but do not consider the adaption of existing mappings to changes beyond preserving consistency.

A metric for inconsistencies in mappings and a method for finding subsets of coherent mappings have been proposed by Meilicke [13]. Ji et al. [11] have demonstrated a tool to preserve the consistency and coherence of ontologies connected through mappings as a part of the NeOn project. While these approaches are able to prevent evolution to cause errors in the ontology networks, they are not able to adapt the mappings to new knowledge, since inconsistencies and incoherencies are corrected by removing mapping entries.

Early work on the mapping adaption problem in the context of ontologies has been performed [12]. This work is focussed on finding which changes are relevant to parts of the mapping. The mappings are formulated as queries. Groß et al. [8] have shown that some techniques from schema mapping adaption can also be applied for ontologies. In their work, they have applied ideas similar to the incremental approach and the mapping composition approach from schema mapping adaption. Regarding precision and recall of the different approaches, the iterative approach outperformed the mapping composition approach, when the composition is constructed from automatically generated mappings between old and new ontology versions. This approach only covers changes in concepts. Structural change or changed relations as well as the impact on the semantics of ontologies are not taken into account. A recent survey concludes that for the field of biomedical ontologies, no approach for "formal axiom-based-ontologies" exist and current approaches only consider simple *is-a* and *part-of* hierarchies [9].

An overview of the current state of the art in mapping adaption in Knowledge Organization Systems (KOS) in general is presented by Dos Reis et al. [5]. The state of the art approaches are divided into mapping revision, that finds contradictions in mapped ontologies and tries to repair them, mapping recalculation, that focusses on a new calculation of mappings when ontologies change, mapping adaption, that changes existing mappings according to the changes in the ontologies, and mapping representation, that deals with representing mappings in a user-friendly way. The survey states that a combination of mapping adaption and mapping representation is the most promising direction for future work. Apart from the state of the art, important research questions are identified. One of those research questions is how to take the semantics of changes into account and reuse the knowledge from the old mapping. Since the semantics of ontologies are defined by inferences that can be drawn from them, the question of taking the semantic changes into account can be restated as how to exploit the *changes of the inferences* of mapped ontologies when they evolve to adapt an existing mapping to the new situation. This issue lies at the core of my research. A hypothesis relevant to this work is that helpful information can be gained by considering changes of the inferences of mapped ontologies that can not be found using the discussed approaches. For example, in ontologies that use relations or classes, that inherit from standard ontology language elements, existing approaches would not be able to find a connection because the specialized elements are unknown to them. Therefore rule-based iterative approaches and compositional approaches not considering inferences would not be able to determine all required information for the mapping adaption in this setting. While a rule could be used to account for each specific case, a more general solution is desirable in this and similar situations.

4   Proposed Approach

The main research question of the doctoral work described in this paper is "*How can the reuse of mappings be ensured, when ontologies evolve?*". This work will address the research gaps mentioned in Sect. 3, namely the currently missing attention for changing semantics after ontology changes. Since the semantics of models in the semantic web are defined by inferences that can be drawn from them and it is considered good modeling practice to base the structure of models on desired inferences [2], the mapping adaption approach is based on inferences before and after the ontology change. These inferences are compared to find previously unrecognised differences. To find out which of these differences are intentional, a classification mechanism is required. This classification mechanism is based on rules, that define intended differences based on changes that occurred in the ontologies. Finding these rules is an important part of my doctoral work.

The differences between the inferences before and after the change are called the *inference diff set*. $\inf(M_{\mathcal{O}_1,\mathcal{O}_2}, \mathcal{O}_1, \mathcal{O}_2)$ denotes the set of inferences from the Mapping M, and the ontologies $\mathcal{O}_1$ and $\mathcal{O}_2$. The diff set is a set of differences in the inferences (additional inferred statements and removed inferred statements) computed by comparing $\inf(M_{\mathcal{O}_1,\mathcal{O}_2}, \mathcal{O}_1, \mathcal{O}_2)$ and the inference set $\inf(M_{\mathcal{O}_1,\mathcal{O}_2}, \mathcal{O}_1, \mathcal{O}_2')$. This will lead to a set of additions and deletions of inferred triples denoted by Δ_{inf}. For the automation of this process, performance needs to be taken into account. Since recomputing all inferences from the ontologies is time consuming, it is reasonable to only consider specific regions of the ontologies that are actually affected by the change. Ideally, these regions should be self-contained ontology fragments. If self-contained fragments are too big, this could be done by setting limits for the distance from changed nodes in the ontology.

For each entry in the inference diff set Δ_{inf} it can be checked, whether the entry is in this set intentionally or because it is accidentally introduced as a byproduct of a change. This can be computed based on rules depending on ontology change patterns. Using this approach, we can compute a set of intentional entries in Δ_{inf} for certain types of changes in the ontologies. For elements that are not intentionally part of Δ_{inf}, the mapping statements of $M_{\mathcal{O}_1,\mathcal{O}_2}$ have to be revised. New mapping statements resulting in these inferences can be added or old statements that lead to unwanted inferences can be removed. This would eventually result in a new set of mapping statements $M_{\mathcal{O}_1,\mathcal{O}_2'}'$.

To find the rules that classify whether a change is intentional or not, a change pattern model needs to be developed. The granularity of this model is important to represent ontology changes adequately. For example, a concept update in an ontology can be seen as a combination of removal and addition operations. Modeling this change as a concept update is more useful to an update of the mapping, while sequential changes (removal and addition) would result in information loss. This is why the representation of changes has a high impact on updating mappings.

5   Methodology for Implementation and Evaluation

To implement the ideas from Sect. 4, an iterative, prototype-based approach will be used. The prototype will be evaluated quantitatively based on benchmarks as well as real-world data. The first concept that needs to be validated is the idea of finding an inference diff set, which will be examined using a prototype. A change pattern model will be developed by studying relevant approaches to ontology change patterns from the literature and from data in the application domain of this doctoral work. The prototype will be extended to incorporate rules based on the change model to classify elements of the inference diff set Δ_{inf} and to propose changes to the ontology mapping. To examine the effectiveness of this approach, a quantitative evaluation of precision and recall regarding changes in real world ontology mappings will be conducted.

The first prototype iteration will incorporate a method to compute inference diff sets from given mapped ontologies when they change over time. Existing mappings from ontology mapping contests by the Ontology Alignment Evaluation Initiative [14] can be used as test data to verify this approach. These mappings and ontologies are freely available and the ontologies are relatively small, which makes debugging and manually changing the ontologies easier.

To examine the issue of creating a comprehensive change pattern model, I will investigate existing ontology change management systems, that already have a change pattern model (e.g. COnto-Diff [10]) and evaluate what kinds of changes are represented in these systems. Also, the results of similar studies from other fields (e.g. [18]) will be incorporated. These models will then be evaluated for their applicability to ontology mapping adaption. Additionally, I will perform case studies for the change model in mapped ontologies used in the software development process to prove the relevance of found change models. Finding a change pattern model will also incorporate defining rules with intended entries in Δ_{inf} for each change pattern. To use the change pattern model, the prototype will be expanded to incorporate the model, which will lead to a second prototype iteration. At first, the tool will allow the classification of changes by hand, so the tool can apply the intended consequences to inference changes, show which entries in Δ_{inf} have been added unintentionally and propose changes to the ontology mapping. An automation of the change classification is very dependent on the complexity of the change model that will be developed. A pattern matching approach could be used to classify changes, but more complex machine learning methods, as well as user interaction, might also be relevant.

The resulting tool can be quantitatively evaluated similar to the approach from Groß [8]. This work examines existing ontology and mapping histories and recomputes the mappings by applying their approach. Since the correct mappings are part of the history, these results can be compared via precision, recall and f-measure calculation. For a qualitative evaluation, case studies in the respective fields will be performed, so that the process will be tested for example ontologies.

6   Results

Currently, the first phase of the plan described in Sect. 5, an implementation of a prototype to compute inference difference sets has begun. The prototype has been implemented using the Jena API [3] in Scala. The tool takes as input two domain ontologies, a mapping between them and a modified domain ontology. It then computes the inference diff set and returns this result.

The prototype can also be fed with a maximum distance of nodes to the changed ontology node. This distance is used to determine which parts of the ontology are used for computing the inferences. The prototype has used the examples used in the Open Alignment Evaluation Initiative contests, namely the "Conferences" example. The changed node has to be specified manually. Using the distance option naturally accelerates the reasoning process for the examples. Even for distances of only three hops in the ontology graph, the result is the same compared to using the full ontology graph.

7   Conclusions and Future Work

This paper presents the approach and research questions of my doctoral work. The central part of the approach is a systematic process that takes inferences of ontologies connected through ontology mappings into account. Besides the approach, the methodology and the current state of the work is presented.

As a next step, the approach will be refined by applying it to ontologies from application domains, namely ontologies in the software development process, and evaluating it based on the described evaluation approach. Comparing my approach to other approaches from the field will provide additional evidence on the effectiveness of the compared approaches and will therefore improve the understanding of the ontology mapping adaption problem in general.

Acknowledgements. I would like to thank my advisors, Bodo Igler from RheinMain University of Applied Sciences and Uwe Brinkschulte from Goethe University Frankfurt am Main.

References

1. Alchourrón, C.E., Gärdenfors, P., Makinson, D.: On the logic of theory change: partial meet contraction and revision functions. J. Symbolic Logic **50**(2), 510–530 (1985)
2. Allemang, D., Hendler, J.: Semantic Web for the Working Ontologist: Effective Modeling in RDFS and OWL, 2nd edn. Morgan Kaufmann Publishers Inc., San Francisco (2011)
3. Apache Foundation: Apache Jena (2016). https://jena.apache.org/. Accessed Sept 2016
4. De Graaf, K.A., Tang, A., Liang, P., Van Vliet, H.: Ontology-based software architecture documentation. In: Proceedings of the 2012 Joint Working Conference on Software Architecture and 6th European Conference on Software Architecture, WICSA/ECSA 2012, pp. 121–130 (2012)
5. Dos Reis, J.C., Pruski, C., Reynaud-Delaître, C.: State-of-the-art on mapping maintenance and challenges towards a fully automatic approach. Expert Syst. Appl. **42**(3), 1465–1478 (2015)
6. Euzenat, J., Shvaiko, P.: Ontology Matching. Springer, Heidelberg (2007)
7. Filipowska, A., Hepp, M., Kaczmarek, M., Markovic, I.: Organisational ontology framework for semantic business process management. In: Abramowicz, W. (ed.) BIS 2009. LNBIP, vol. 21, pp. 1–12. Springer, Heidelberg (2009). doi:10.1007/978-3-642-01190-0_1

8. Groß, A., Dos Reis, J.C., Hartung, M., Pruski, C., Rahm, E.: Semi-automatic adaptation of mappings between life science ontologies. In: Baker, C.J.O., Butler, G., Jurisica, I. (eds.) DILS 2013. LNCS, vol. 7970, pp. 90–104. Springer, Heidelberg (2013). doi:10.1007/978-3-642-39437-9_8

9. Groß, A., Pruski, C., Rahm, E.: Evolution of biomedical ontologies and mappings: overview of recent approaches. Comput. Struct. Biotechnol. J. **14**, 1–8 (2016)

10. Hartung, M., Groß, A., Rahm, E.: COnto-Diff: generation of complex evolution mappings for life science ontologies. J. Biomed. Inf. **46**(1), 15–32 (2013)

11. Ji, Q., Haase, P., Qi, G., Hitzler, P., Stadtmüller, S.: RaDON — Repair and Diagnosis in Ontology Networks. In: Aroyo, L., Traverso, P., Ciravegna, F., Cimiano, P., Heath, T., Hyvönen, E., Mizoguchi, R., Oren, E., Sabou, M., Simperl, E. (eds.) ESWC 2009. LNCS, vol. 5554, pp. 863–867. Springer, Heidelberg (2009). doi:10.1007/978-3-642-02121-3_71

12. Klein, M., Stuckenschmidt, H.: Evolution management for interconnected ontologies. In: Workshop on Semantic Integration at ISWC (2003)

13. Meilicke, C.: Alignment incoherence in ontology matching. Ph.D. thesis, Universitätsbibliothek Mannheim (2011)

14. Open Alignment Evaluation Initiative: OAEI Home page (2016). http://oaei.ontologymatching.org/. Accessed Sept 2016

15. Paulheim, H., Probst, F.: A formal ontology on user interfaces. Yet another user interface description Language? In: Proceedings of the 2nd Workshop on Semantic Models for Adaptive Interactive Systems in conjunction with the 2011 International Conference On Intelligent User Interfaces (IUI 2011) (2011)

16. Ribeiro, M.M., Wassermann, R.: First steps towards revising ontologies. In: Proceedings of WONRO 2006, pp. 1–11 (2006)

17. Shvaiko, P., Euzenat, J.: Ontology matching: state of the art and future challenges. IEEE Trans. Knowl. Data Eng. **25**(10), 158–176 (2013)

18. Thor, A., Hartung, M., Groß, A., Kirsten, T., Rahm, E.: An evolution-based approach for assessing ontology mappings - a case study in the life sciences. In: Proceedings of Conference of the Business, Technology and Web (BTW), pp. 277–286 (2009)

19. Velegrakis, Y., Miller, R.J., Popa, L.: Mapping adaptation under evolving schemas. In: Proceedings of the 29th International Conference on Very Large Data Bases, VLDB 2003, vol. 29, pp. 584–595 (2003)

20. Yu, C., Popa, L.: Semantic adaptation of schema mappings when schemas evolve. In: Very Large Data Bases, pp. 1006–1017 (2005)

21. Zablith, F., Antoniou, G., D'Aquin, M., Flouris, G., Kondylakis, H., Motta, E., Plexousakis, D., Sabou, M.: Ontology evolution: a process-centric survey. Knowl. Eng. Rev. **30**(01), 45–75 (2013)

Modeling, Exploring and Recommending Music in Its Complexity

Pasquale Lisena[✉]

EURECOM, Sophia Antipolis, France
pasquale.lisena@eurecom.fr

Abstract. Knowledge models that are currently in-use for describing music metadata are insufficient to express the wealth of complex information about creative works, performances, publications, authors and performers. In this thesis, we aim to propose a method for structuring the music information coming from heterogeneous librarian repositories. In particular, we research and design an appropriate music ontology based on existing models and controlled vocabularies and we implement tools for converting and visualizing the metadata. Moreover, we research how this data can be consumed by end-users, through the development of a web application for exploring the data. We ultimately aim to develop a recommendation system that takes advantage of the richness of the data.

Keywords: Ontology · FRBRoo · Music metadata · Schema.org · Recommender system

1 Problem Statement

Music metadata can be very complex. Metadata describing a well-known masterpiece such as the *Moonlight Sonata* can describe its composition by Beethoven, its scores in the handmade or printed version, the interpretations by pianists, the orchestrations and arrangements. Performances, recordings, music albums can also be described. Numerous actors are involved in this media production chain: composers, performers with different roles, conductors, etc.

Online musical content holders offer a very simplified version of this information, focused on the track as the atomic unit, the artist as the unique carrier of the authorship, and presence in the same album as unique possible relationship between tracks. While a simplified version of the metadata is enough for commercial purpose, the whole complexity of the music information opens up new possibilities for advanced search, visualization, and recommendation strategies.

Libraries, in contrast, have more structured information, often represented in specialized formats such as MARC[1]. The limit of this approach is that developers are tied to these formats with non-explicit semantics. Moreover, each institution hosts data in its repositories, often using a particular dialect of MARC. As a

---

[1] https://www.loc.gov/marc/.

© Springer International Publishing AG 2017
P. Ciancarini et al. (Eds.): EKAW 2016 Satellite Events, LNAI 10180, pp. 264–268, 2017.
DOI: 10.1007/978-3-319-58694-6_41

result, the music metadata has currently no chances of reconciliation and inter-connection. The benefits about moving from MARC to an RDF-based solution consist in the interoperability and the integration among libraries and with third part actors, with the possibility of realizing smart federated search [2].

This thesis aims to propose a new model for representing music metadata in its full complexity and design, implement and evaluate novel applications that use this metadata with the purpose of exploring and recommending music.

This research is being developed in the context of the DOREMUS project[2] [1], in which three leading cultural institutes in France — the BnF (Bibliothèque Nationale de France), the Philharmonie de Paris and Radio France — join forces with companies and academic institutions in order to make the music knowledge in their catalogs available and re-usable on the web of data.

2   Proposed Approach

Three different challenges represent the ambitious goals of this research thesis.

The first challenge is to find an appropriate ontology model for capturing the richness of music information, taking into account all its components. The data in MARC format from the source institutions will be converted independently in RDF. After that, a reconciliation process should be realized. sameAs links should be found on the resources from the different institutions that describe the same work and let them converge in a unique (virtual) graph containing the whole information at our disposal. This should be realized by identifying the features that enable to disambiguate resources. Controlled vocabularies must be used for describing features as keys, genres, etc., and disambiguating their values.

The second challenge consists in providing a simplified version of this struc-tured metadata, tailored to be consumed by search engines and third-part appli-cations. Because of its central role in the web, we will provide mappings of this ontology into Schema.org. Some research questions are: *1. which information should be presented in this simplified version and which one is considered too complex? 2. which methods should be used for realizing this simplification?*

Finally, our aim is to demonstrate the benefits that this data can produce once being consumed and displayed to the end-user. We consider taking the user into account as a crucial requirement for the design of these systems. In this context, an application will be developed that consists in two parts. First, an exploratory search engine for musical data will be designed and developed. *1. Can the knowledge model simplification operated in the second challenge be used to improve the user experience? 2. How complex concepts and relationship should be displayed to the end-user?*

On top of the exploratory interface, we will build a recommendation engine. *1. Are the recommendation systems currently in-use for music still valid with this complexity? 2. Is the rich model better than the simplified (*Schema.org-based*) one for feeding the recommendation?*

---

[2] http://www.doremus.org.

3   State of the Art

Musical Data Representation. The description of music is historically connected to catalog information models, among which FRBR is one of the most popular. FRBR and CIDOC-CRM, an ontology for describing museum information, have been harmonized in the FRBRoo ontology for describing arts [4]. This model considers a cultural entity as the combination of the Work that is realised into an Expression through an Event creation. Among music specific ontologies, the most popular is Music Ontology [10]. The experiences about converting data from MARC to RDF include MARiMbA [5], that uses the FRBR model and manages also the interlinking, loading and visualization of the data.

Music Recommendation. Music recommendation is an active and popular research field. Most of the important actors in online music provide some sort of recommendations, often making use of collaborative recommendation algorithms. The need to discover novelty in results requires the use of approaches like the Long Tail curve for also including less popular works in recommendation [3]. A feature-combination hybrid approach is presented in [9]: music datasets are semantically enriched through external graphs and combined with collaborative features for providing better results in terms of both accuracy and novelty. The approaches completely based on the knowledge of the model are less common, like a recommendation system driven by the nearby places of interests [6].

4   Methodology

Different parts of this thesis are currently carried out in parallel, in order to have in mind the final goals during the development of the various strategies.

DOREMUS Ontology. The DOREMUS ontology[3] is being developed as an extension of the FRBRoo model. This choice comes from two reasons: a fine granularity of description granted by the triplet pattern, and the possibility to interoperate with systems dealing with any kind of cultural data. A modeling group of experts in music cataloging and knowledge representation, is currently working on the ontology, that currently defines 46 classes and 144 properties.

Controlled Vocabularies. Different kind of vocabularies are needed for describing music: some are already available on the web (like medium of performance or musical genres), while others do not exist (like the types of derivation). Not all of them are published in a suitable format for the Web of Data, and there is often little interconnection between vocabularies. Our effort is to publish SKOS representation for each controlled vocabulary, with appropriate relationships and mappings between the different sources.

---

[3] http://data.doremus.org/ontology.

Simplification Through Schema.org. Schema.org contains different types for describing music: CreativeWork, MusicComposition, MusicRecording, Music-Group, etc. Passing from FRBRoo to the Schema.org means mapping the concept expressed in a complex ontology to a simpler one. We have proposed a method composed of a series of recipes that enables to perform this operation based on the observation of the graph and the individuation of the relative Schema.org types through similarity criteria [8].

Recommendation. Our starting point for the study of a recommendation algorithm consists in two human-made collections: playlists from Radio France web radios and concert programs from Philharmonie of Paris, realized by an editorial action made by experts and aiming at discovering novel works. In parallel, we will consider the leading music providers: we are developing ARyTREx[4], a web application that aims to simulate the recommendations made by Spotify and Last.fm given a seed artist or track and using their respective APIs. The application enables to collect recommendation paths that can feed a semi-supervised machine learning algorithm. One goal will be to identify which features are the most important in the recommendation. Programs, playlists and third-parts recommendation results will play a key role in the evaluation of the algorithm, that can be iteratively corrected by the comparison between its output and these collections. Precision and recall metrics will be used to measure this accuracy.

5 Preliminary Results

We have already developed two prototypes, both as open source software [7].

MARC2RDF[5] is a prototype for converting the bibliographic records from UNIMARC and INTERMARC to an RDF graph following the DOREMUS model. The conversion relies on explicit expert-defined transfer rules (or mappings), that define for each property/class in the DOREMUS ontology where to find the relative information in the MARC file. The software contains also a *string2uri* component, that performs an automatic mapping of string literals to URIs coming from controlled vocabularies.

A prototype of an exploratory search engine named OVERTURE[6] is under development. The challenge is in giving to the final user a complete vision on the data of each work, performance, score, recording and letting him/her to understand how they are connected to each other. The exploratory interface is available at http://overture.doremus.org.

6 Conclusion and Future Work

Representing the information about classical music is a complex activity, that involves different sub tasks. During this research, we want to express the musical metadata from their current MARC format in a suitable ontology that can

---

[4] https://github.com/fernanev/ARyTREx.
[5] https://github.com/DOREMUS-ANR/marc2rdf.
[6] https://github.com/DOREMUS-ANR/overture.

preserve their complexity. From that, a simplification through Schema.org will guide the realization of an exploratory interface, while the richness of the model will fully support the development of a recommender system.

We are publishing a large DOREMUS dataset composed of the ontology, controlled vocabularies and data about classical and jazz works, accessible at data.doremus.org/sparql. Disambiguation of data and interlinking between resources should be performed in the context of controlled vocabularies. The data coming from the conversion of partner institutions will be interlinked, so that different descriptions of the same work go enriching each other. Then, this research will go deep in the data presentation and recommendation.

Acknowledgments. I would like to thank my supervisor Raphaël Troncy for his ongoing support in. This work has been partially supported by the French National Research Agency (ANR) within the DOREMUS Project, under grant number ANR-14-CE24-0020.

References

1. Achichi, M., Bailly, R., Cecconi, C., Destandau, M., Todorov, K., Troncy, R.: DOREMUS: doing reusable musical data. In: 14th International Semantic Web Conference (ISWC) (2015)
2. Byrne, G., Goddard, L.: The strongest link: libraries and linked data. D-Lib Mag. **16**(11), 5 (2010)
3. Celma, Ò.: Music recommendation and discovery in the long tail. Ph.D. thesis, Universitat Pompeu Fabra (2009)
4. Doerr, M., Bekiari, C., LeBoeuf, P.: FRBRoo: a conceptual model for performing arts. In: CIDOC Annual Conference, pp. 6–18 (2008)
5. Greenberg, E., Gema Bueno de la Fuente, J., Vila-Suero, D., Gómez-Pérez, A.: datos.bne.es and MARiMbA: an insight into library linked data. Library hi Tech **31**(4), 575–601 (2013)
6. Kaminskas, M., Fernández-Tobías, I., Ricci, F., Cantador, I.: Knowledge-based music retrieval for places of interest. In: 2nd International ACM Workshop on Music Information Retrieval with User-Centered and Multimodal Strategies, pp. 19–24 (2012)
7. Lisena, P., Achichi, M., Fernandez, E., Todorov, K., Troncy, R.: Exploring linked classical music catalogs with OVERTURE. In: 15th International Semantic Web Conference (ISWC) (2016)
8. Lisena, P., Troncy, R.: DOREMUS to Schema.org: mapping a complex vocabulary to a simpler one. In: 20th International Conference on Knowledge Engineering and Knowledge Management (EKAW) (2016)
9. Ostuni, V., Oramas, S., Di Noia, T., Serra, X., Di Sciascio, E.: Sound and music recommendation with knowledge graphs. ACM Trans. Intell. Syst. Technol. (TIST) **8**(2), 21:1–21:21 (2016). doi:10.1145/2926718
10. Raimond, Y., Abdallah, S., Sandler, M., Giasson, F.: The music ontology. In: 15th International Conference on Music Information Retrieval (ISMIR), vol. 422 (2007)

Metropo-Lifeline: Participatory Description and Analysis of the Migration of Residents Within a Metropolitan Area

David Noël[✉]

Univ. Grenoble Alpes, LIG, 38000 Grenoble, France
david.noel@imag.fr

Abstract. This multidisciplinary research aims to allow urban planners and decision makers to better understand urban and/or peri-urban migrations in order to adjust their decisions. The more specific question concerns the residential trajectories of inhabitants (i.e. the succession of their residential choices in time and space). To better understand the circumstances that lead people to move, we propose a model that takes into account the different aspects of the individuals lives (family professional, spare-time activities, etc.) and allows to elicit explanatory factors between life events (e.g. a move due to a birth to come). Our approach also leads us to design an innovative participatory software to collect and analyze descriptive and localized data on the residential trajectories between different territories of a metropolitan area.

Keywords: Residential trajectory · Life trajectory · Semantic trajectory modeling · Participatory description

1 Introduction

Residential choices of metropolitan areas inhabitants are strongly influenced by the various events that arise in their personal lives. For instance, in general, couples with children prefer to live in peri-urban areas [1] or they simply cannot afford to do another way. This exodus of working couples with (young) children to peri-urban areas requires that cities of the peripheries provide inhabitants with the necessary amenities (for instance, nurseries and schools) with appropriate opening hours. It also increases the time workers need to commute from home to work, which generates pollution and traffic congestion.

Providing people with greater opportunity to live in city centers is an important issue for decision makers and urban planning experts who need to acquire a better knowledge of residential dynamics. In particular, in order to adapt their decisions, they try to better understand what is at stake for an individual or a household in the process of choosing a residence. If overall patterns of residential migrations are already known, there is still a lack of qualitative data, collected from individuals, on the reasons underlying the residential choices.

The living conditions, especially the residence and its neighborhood, as well as other elements linked to the economic conditions must be considered [4]. However, underlying reasons can only be provided from a comprehensive approach of individuals' life

© Springer International Publishing AG 2017
P. Ciancarini et al. (Eds.): EKAW 2016 Satellite Events, LNAI 10180, pp. 269–277, 2017.
DOI: 10.1007/978-3-319-58694-6_42

over time. Then, in order to study the *residential trajectory* someone (the succession of its residential choices in time and space), we should consider it as an integrated part of a broader vision of her/his life trajectory that includes decisive aspects of life (residential but also familial, professional, etc.) and the links between theses aspects. The idea here is to understand the interactions of the different dimensions of life trajectories.

The collect of such data about people life trajectory is challenging but can benefit from the rise of crowdsourcing and Volunteered Geographic Information. Information generated by the *citizen as sensor* [5] can be used to improve public services or to adapt urban planning decisions. For example applications like FixMyCity5 allows citizens report local problems, (degradation…) in order to help services of the municipality. This illustrates the potential of a participatory tool in the field of urban planning [14]. In our context, the final step will be to offer to urban planners and decision makers an innovative tool that allows them, from real cases collected through a participatory approach, to better understand urban and/or peri-urban migrations and specifically the reasons for residential choices. This requires addressing different issues, from a conceptual level (to represent peoples' life trajectory in a way adapted to the use which will be made of these data), to an operational level to both collect and then restitute useful data. The work we present here mainly deals with the conceptual issues.

This paper is organized as follows. Section 2 presents the problems of studying residential choices over time. Section 3 introduces briefly the state of the art. Section 4 is centered on our approach including methodology and preliminarily results. We draw a conclusion and perspectives for this work in Sect. 5.

2 Issues to Address

Reasons for residential choice can only be provided from a comprehensive approach to individuals' life trajectories that takes into account familial or professional circumstances, but others aspects such as leisure could also be decisive in the choice of people. A model for the life trajectory has therefore to be flexible enough in order to characterize what is relevant for each individual. Furthermore the model has to include explanatory factors in order to represent not only the different life events but also the connections that can be established between them. For example, a move may be explained by a change in the professional life (career development, job loss, etc.). Beyond this trivial example, one issue is the representation of the combination of several explanatory factors of various kinds (a change of job may have made a move possible, the next arrival of a baby may have made it imperative). These significant events have also to be linked with characteristics of the residence, the neighborhood and external circumstances (e.g. economic conditions…) to be fully understand as explanatory factors.

Beyond the proposition of a life trajectory model, other issues are related with data collection. Regarding data collection relying on a participatory approach, contributor oversight (mistakes or omissions) is a risk, especially when he/she has to remember past events. The contributor could then be lead to consider a life event at a high scale (for example "*in the 80s*") or with a lack of precision ("*about 1990*"). This refers to the question of the granularity of the information which is fundamental in moving object modeling [10]. In the case of temporal granularity, each piece of information should be given at an ideal *chronon* (i.e. the smallest temporal unit considered [9]), but we must be able to manage higher granularities. For example, the day is an adequate *chronon* for some life event (a wedding, a birthdate, etc.) but individuals will surely report a life event at the scale of the year or even the decade. A similar reasoning is applied to geographic information [12]. Another issue is to bring individuals to contribute although they can be reluctant to share private data. In addition to ensure anonymity of contributors, we have to propose them a playful and useful environment. Another possible direction here is to harvest social networks to initiate the data collection process.

Regarding data exploitation, in the dialogue with the future users (i.e. decision makers and urban planning experts) about the functionalities they expect, we pay a special attention to the information granularity level. Indeed the granularity level at which an information is presented may reveal different phenomena. It is therefore important to allow spatial and temporal exploration and analysis at various temporal and spatial scales. We focus on both the proposition of performing algorithms for data aggregation and the proposition of appropriate interfaces for data exploration and restitution.

3 State of the Art

3.1 Studying Residential Choice

Conditions of residential choices have been well documented in the last decades in particular by sociologists and demographers. The use of a biographic perspective, meaning the fact to consider residential choice in relation with the *life course* and the various life cycles (professional, familial...) has contributed to enrich the modeling of this notion [2]. In this context, first works focused specifically on highlighting relationships between the multiple events that arise in the life of individuals. In many cases, the principle is to model the imbalance resulting by event happened in the life of individuals (job change, child birth...) and to consider that this imbalance causes a residential move when it reaches to a certain point [10]. More recently, other works remind that even if the biography of individuals is often considered as a sequence of linked events, the importance residential stability periods is not to be underestimated [3] because life of individuals contain at the end relatively few moves, effective or even just considered. For a better understanding of these residential stability periods, researchers highlight the importance to consider aspirations of individuals, in order to put in relation an effective trajectory and a projection towards the future of the individuals [3].

3.2 Trajectory Modeling

Semantic Trajectory Modeling. Following Hägerstraand and the "time-geography" [6], researchers have modelled the spatio-temporal trajectories through periods of trips and periods of activities, or through moves and stops [13]. These structures are then used to add information about the trajectory, like the transportation mode during a trip, or the nature of an activity. These authors propose especially solutions to enrich raw GPS data indicating the path of a moving object, which result in trajectories encompassing useful information in their description. Some of these models have been integrated in the semantic web [7] but they are not suitable for characterizing life trajectories because they focus on the spatial component of trajectories which is not is not essential in a life trajectory.

Life Trajectory Modeling. The life trajectories are a particular type of semantic trajectories. The work conducted by Marius Thériault is a reference in the field of the life trajectory modeling. The space-time model for the analysis of life trajectories [15] is based on three different trajectories - residential, family and professional. Each of these trajectories is conceptually modeled by *episodes* - stable statutes during a time interval - and *events* altering one or more of these statutes. Later, the model has been modified in order to determine the likelihood of an event occurring under certain conditions [16]. The conditions of the residential choices are also addressed by [17] with a strong emphasis on the role of job changes occurring in the professional life of individuals. The models proposed in these works are centered on the temporal aspect of residential choices: the focus is more on *why* people have moved at some point in their life than on *why they have chosen* the place where they have moved to. Finally these models don't have enough flexibility/genericity to allow characterization of the multiple parts of the life trajectory that can be different for each individuals and according to the application case.

3.3 Characterize Space and Time

Modeling life trajectories in order to study residential choice requires to characterize space and time and particularly the urban space. There are widely known ontologies of temporal or spatial concepts that we propose to reuse in our proposal. For example the OWL-Time ontology[1] is a W3C standard which can be used in the development of a trajectory ontology [7]. The GEOSparql ontology[2], a standard proposed by the Open Geospatial Consortium (OGC), allow to characterize spatial concepts. This ontology enables the use on the GEOSparql Query Language for querying geospatial data.

Useful ontologies have also been developed to characterize the territory and particularly the urban space. For example OSMonto[3] is an ontology extracted from OpenStreetMap (OSM) tags. The use of this ontology allows to access OSM data (location of amenities, roads network...) in a semantic web context. Other ontologies characterize topological entities[4] or French administrative divisions[5].

---

[1] https://www.w3.org/TR/2016/WD-owl-time-20160712/.

[2] http://www.opengeospatial.org/standards/geosparql.

[3] http://wiki.openstreetmap.org/wiki/OSMonto.

[4] http://data.ign.fr/def/topo/20140416.en.htm.

[5] http://data.ign.fr/def/geofla/20140822.en.htm.

4   Proposed Approach

4.1   Overview

A schematic representation of our approach is given in Fig. 1. We adopt a two-step approach. The first step is to create an original life trajectory model that fits our requirements and the second is to design an environment that allows data collection and data exploitation. At each step we use the possibilities offered by the semantic web as data source but also as a source of knowledge allowing multiscale analysis and data inference thanks to available standard ontologies.

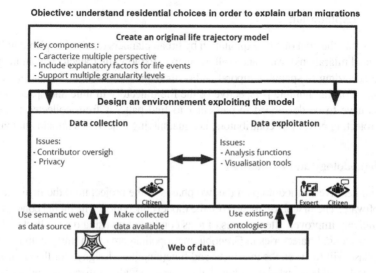

Fig. 1. Overview of the proposed approach

4.2   Life Trajectory Modeling

We propose to model the concept of life trajectory following the three points cited previously. First the model is generic and support the taking into account of any thematic that reveals itself relevant in urban migration. Secondly the model includes explanatory factors for residential choices and more widely for life event. Finally multiple granularity levels are supported to conform to the spatio-temporal nature of the data we expect to collect. A study has been conducted during which 30 individuals were interviewed on their life trajectories. The collected data have been used to test the expressive power of our model and to improve it. In addition to the model, a set of requests useful for data exploitation has been proposed and tested (cf. Sect. 4.6).

4.3 Data Collection

A particular effort is made to facilitate the data collection from contributor: spatio-temporal interfaces are proposed to give participants contextual clues and functionalities to be as precise as possible in their information delivery. Contributors are invited to provide explanation about residential events and typologies are proposed for explanatory factors elicitation. Our approach to motivate a participant is to offer her/him an environment in which she/he contributes but also has a pleasant feedback. The contributors can therefore visualize and explore their own data in space and time in a playful interface. They also have access to statistics of a general interest on the project.

4.4 Data Exploitation

With regards to the goal of data exploitation by urban planners (analyze explanatory factors of residential migrations), we want to allow them to explore life trajectory in an interactive interface according to spatial, temporal and/or thematic criteria. The interfaces allow the visualization of the multiple perspective of the life trajectory in time and space. Analysis functions have to be developed in particular to find patterns from collected trajectories (while protecting privacy of contributors, i.e. guarantying trajectories remain anonymous).

4.5 Methodology and Evaluation

We use an iterative methodology in the two phases of the project until the final evaluation of the software. The first phase focused on the model conception. The model conception has been iteratively improved after a series of tests conducted by using data collected during a study on residential trajectories in Grenoble metropolitan area with thirty individuals. The second phase will focus on the interfaces and functionalities designed for the environment. First user tests will be assisted, meaning that the user (including citizens and urban planning experts) is assisted in the use of the prototype before they test it without any assistance (to simulate an online version of the tool). A questionary will be proposed to collect users' qualitative feedback.

4.6 Preliminary Results

We have proposed a new life trajectory model (cf. section 4.2) that enables the representation of any thematic aspect that "makes life": professional, family, residential, leisure, etc. A simplified view of the semantic trajectory model is presented in Fig. 2. This model rests on a semantic trajectory model composed of *Episodes,* which describe the *stable state* of a person when observed or considered from a given viewpoint (residential, professional…). It is also composed of *Events* that mark the transition from one stable state to another (for example a move in the residential viewpoint). Explanatory factors are constituted of weighted links between events (*isExplainedby*) associated with relevant information (for example an Event *move* is linked to an Event *child birth* and a relevant information brought by the individual is "we needed one more room"). This model and our first results are presented in [11].

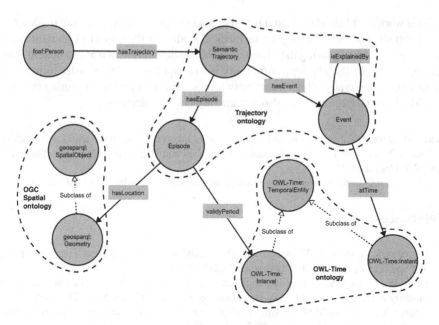

Fig. 2. Simplified overview of the semantic trajectory model (View with VOWL: http:// vowl.visualdataweb.org)

The model is connected to some of the standard ontologies presented in Sect. 3. The use of OWL-Time and OGC Spatial ontologies allow to store and then exploit spatial and temporal data at different levels of granularity using the GEOSparql query language. The model is also linked to other ontologies such as the topological ontology and the administrative division ontology of the French National Geographical Institute. The model also includes explanatory factors that provide additional information useful to understand the choice people made in their lives. The model has been evaluated using real story telling collected through a survey of 30 respondents.

5   Conclusions and Future Work

Our work deals with the modeling of life trajectories with an emphasis on residential choices so to help decision-making in the field of cities development. Designing a tool, relying on a participatory approach for data collection, and offering ad hoc functionalities for spatio-temporal data exploration, has also guided our proposition.

We show that there is no available model in the literature that fit our requirements. Based on this observation, we proposed an approach based on a generic life trajectory modeling including explanatory factors and multiple granularity level connected to the semantic web, as a data source and as source of knowledge. We have also started to investigate an approach for data collection and data exploitation through dedicated spatio-temporal interfaces.

This work could also be adapted for other application cases in which life trajectories are important. For example long terms medical studies or the impact of tourism experiences on individuals during their lives could be application cases that can benefit from our approach. Finally, another future work is to adapt the model in order to use it to characterize more types of high granularity level semantic trajectories meaning not only individual (life) trajectories but also any moving object trajectories.

Acknowledgments. I would like to thank my thesis supervisors: Jérôme Gensel and Marlène Villanova-Oliver from *University Grenoble Alpes, LIG* and Pierre Le Quéau from *University Grenoble Alpes, PACTE.*
This doctoral thesis benefits from a grant of the French Rhône-Alpes Region Council.

References

1. Bonvalet, C., Laflamme, V., Arbonville, D.: Family and Housing: Recent Trends in France and Southern Europe. Bardwell Press, Oxford (2009)
2. Clark, W., Whiters, S.: Family migration and mobility sequences in the United States: spatial mobility in the context of the life course. Demographic Res. **17**(20), 591–622 (2008)
3. Coulter, R., Van Ham, M.: Following people through time: an analysis of individual residential mobility biographies. Hous. Stud. **28**(7), 1037–1055 (2013)
4. Dieleman, F.M.: Modelling residential mobility; a review of recent trends in research. J. Hous. Built Environ. **16**(3), 249–265 (2001)
5. Goodchild, M.F.: Citizens as sensors: the world of volunteered geography. GeoJournal **69**(4), 211–221 (2007)
6. Hägerstraand, T.: What about people in regional science? Papers in regional science (1970)
7. Hu, Y., Janowicz, K., Carral, D., Scheider, S., Kuhn, W., Berg-Cross, G., Hitzler, P., Dean, M., Kolas, D.: A geo-ontology design pattern for semantic trajectories. In: Tenbrink, T., Stell, J., Galton, A., Wood, Z. (eds.) COSIT 2013. LNCS, vol. 8116, pp. 438–456. Springer, Cham (2013). doi:10.1007/978-3-319-01790-7_24
8. Hornsby, K., Egenhofer, M.J.: Identity-based change: a foundation for spatio-temporal knowledge representation. IJ Geog. Inf. Sci. **14**(3), 207–222 (2000)
9. Jensen, C., Dyreson, C., Böhlen, M., Clifford, J., Elmasri, R., Gadia, S., Grandi, F., Hayes, P., Jajodia, S., Käfer, W., Kline, N.: The consensus glossary of temporal database concepts (1998)
10. Mulder, C.H., Wagner, M.: Migration and marriage in the life course: a method for studying synchronized events. Eur. J. Popul. **9**(1), 55–76 (1993)
11. Noël D., Villanova-Oliver M., Gensel J., Le Quéau P.: Modeling semantic trajectories including multiple viewpoints and explanatory factors: application to life trajectories. In: ACM SIGSPATIAL URBANGIS, Seattle, United States (2015)
12. Openshaw, S., Taylor, P.J.: A million or so correlation coefficients: three experiments on the modifiable areal unit problem. Stat. Appl. Spat. Sci. **21**, 127–144 (1979)
13. Spaccapietra, S., Parent, C., Damiani, M.L., de Macedo, J.A., Porto, F., Vangenot, C.: A conceptual view on trajectories. Data Knowl. Eng. **65**(1), 126–146 (2008)
14. Seeger, C.: The role of facilitated volunteered geographic information in the landscape planning and site design process. GeoJournal **72**(3), 199–213 (2008)

15. Thériault, M., Séguin, A. M., Aubé, Y., Villeneuve, P.Y.: A spatio-temporal data model for analysing personal biographies. In: Tenth International Workshop on Database and Expert Systems Applications, Proceedings, pp. 410–418. IEEE (1999)
16. Thériault, M., Claramunt, C., Séguin, A.M., Villeneuve, P.: Temporal GIS and statistical modelling of personal lifelines. In: Advances in Spatial Data Handling (2002)
17. Vandersmissen, M.H., Séguin, A.M., Thériault, M., Claramunt, C.: Modeling propensity to move after job change using event history analysis and temporal GIS. J. Geogr. Syst. **11**(1), 37–65 (2009)

Facilitating the Management and Analysis of Scholarly Communication Metadata

Sahar Vahdati[✉]

Enterprise Information Systems (EIS), University of Bonn, Bonn, Germany
vahdati@cs.uni-bonn.de

Abstract. Digitizing scholarly communication is a major challenge of our era. In this thesis, we focus particularly on facilitating the digital handling of scholarly communication metadata, i.e. bibliographic data, metadata about scientific events, courseware, projects, organizations etc. We describe these metadata domains and develop representation schemes for semantically representing respective information. We develop a conceptual lifecycle model for facilitating the management of scholarly communication data. Furthermore, we present some concrete strategies and applications for semantically representing and linking bibliographic data, for crowdsourcing and analysing events metadata and for quality assessment of opencourseware and scientific events based on this metadata.

Keywords: Semantic metadata enrichment · Quality assessment · Recommendation services · Scholarly communication · Semantic publishing

1 Problem Statement

Research papers have been the key element of scholarly communication and are crucial for evaluating the academic achievements of researchers. Nowadays, most of the scientific publications are easily available online, but the opportunities of the digital technologies are by far yet exploited. Researchers investigating a new topic still spend a lot of time searching for the most relevant publications, persons or groups inside the respective community. Even for experienced members of a community who may know what sources to explore and where to look for more explanations or what conference's proceedings to read, the relevant information they are interested in is not accessible at a glance. Most of the services are custom implementations serving limited amount of information. In order to provide a better support a semi-automatic crowd-sourcing approach is proposed for systematic and coherent management of scholarly metadata.

The building blocks of this approach are (a) semantic enrichment strategies and (b) quality assessment methods. An additional goal of this work is to develop strategies for identifying similar or related entities with regard to quality and provide recommendations using enriched metadata. We identify three specific research goals: (**RG1**) quality based measure where quality is defined as *fitness for use*; (**RG2**) efficient metadata access; (**RG3**) *better* recommendation services based on semantically enriched metadata. As a result, we propose a logical

© Springer International Publishing AG 2017
P. Ciancarini et al. (Eds.): EKAW 2016 Satellite Events, LNAI 10180, pp. 278–285, 2017.
DOI: 10.1007/978-3-319-58694-6_43

theory that uses semantically enriched metadata as quality assessment metrics for ranking, filtering of the assessment objects in a flexible and user-defined way.

Motivating example. We motivate the problem of recommendation services based on quality assessment with the following scenario. Usually young researchers spend much time finding related information about their topic such as must-read publications of their fields, get quick overview of the state-of-the-art of their topic of interest, and upcoming related scientific events to which they can submit their research contributions. They find suitable events by Web searches, input from mailing lists subscriptions, individual experience, or recommendations from senior researchers. There are already attempts to assist researchers in this task, however, resulting recommendations are often rather superficial and the underlying process neglects the different aspects that are important for authors. Providing recommendation services to researchers and a comprehensive list of criteria while they are searching for information to submit their research contributions is a major need.

2 State of the Art

Several efforts on publishing reusable, machine-readable metadata (i.e. *linked open data*) related to scholarly data such as publication, scientific events, authors and etc., have been motivated quality considerations. The Springer LOD dataset[1] about their conference proceedings (Lecture Notes in Computer Science) serves trust-related questions of stakeholders. Our own ongoing work on extracting linked data from the CEUR-WS.org open access computer science workshop proceedings volumes is also motivated by quality assessment. We run a few dozen of quality-related queries such as "What workshops have changed their parent conference?" against the linked dataset in order to assess the quality of the workshops published at CEUR-WS.org and to validate different information extraction implementations [6,7]. Both the work of Bryl et al. and ours have in common that they lack a systematic, comprehensive definition of quality dimensions. In addition, there are a number of web resources devoted to present information about scientific events and their quality. *WikiCFP*[2], for example, is a repository of calls for papers in science and technology fields. However, WikiCFP neither provides explicit information about the quality of events nor recommendations. Currently, many RDF data are made available, the Semantic Web Dog Food (SWDF) dataset[3] as one of the pioneers and *ScholarlyData*[4] that provides RDF dumps for scientific events. *Google scholar*[5] and *ResearchGate*[6] provides recommendation alerts using person profile on the possible related scientific results but not about other objects such as events, citations and etc.

---

[1] http://lod.springer.com/.

[2] http://www.wikicfp.com/cfp/.

[3] http://data.semanticweb.org/.

[4] http://www.scholarlydata.org/dumps/.

[5] https://scholar.google.de/.

[6] https://www.researchgate.net/.

On the other hand, many domain specific ranked list of conferences and journals are published by different scientific field such as *DBWorld*[7]. Whereas all such lists are created based on limited number of metrics such as acceptance rate.

3   Proposed Approach

We propose a coherent and systematic approach that combines human and computer capabilities to manage scholarly metadata.

3.1   Metadata Domains

The central objects of scholarly communication are publications together with datasets by which scientists exchange knowledge. Scientific events and journals are the main channels of this communication. Organizations including companies, research centers, institutions and etc. are involved as project partners or as responsible of operating data sources. They can appear in different roles such as sponsors, person affiliation and funder of projects that have led to given research results. Metadata domains: Publications, Datasets, Persons, Events, Organizations, Projects, Tools, OCW (OpenCourseWare). In modeling the ontology of scientific communication, we followed the best practices of *reusing* terms from existing ontologies (cf. [8]) and applying ontology *design patterns.* For any further terminology not sufficiently covered by existing ontologies, we defined our own ontology.

Fig. 1. Metadata domains

---

[7] https://research.cs.wisc.edu/dbworld/.

3.2 Metadata Life-Cycle

Managing the metadata of scholarly communication involves a number of stages of the data management life-cycle. Based on the Linked Data life-cycle [2] we propose a life-cycle for scholarly communication metadata. The different stages of the life-cycle include:

Identification: The types of metadata that should be considered for extraction and evaluation varies for every domain. The proper identification of metadata is particularly challenging and important when the metadata is planned to be exploited for determining as quality criteria for the domain objects. An expert or knowledge engineer identifies a set of metadata items and related quality metrics. Any metric has a precise definition by which its exact value can be computed from metadata. We propose a framework for identification and classification of quality indicators that follows the standard terminology of *data quality* research, with the key terms of *category*, *dimension* and *metric*.

Extraction/Transformation: During this stage, data is extracted from unstructured, semi-structured and structured data. The set of identified and defined metadata in the previous life-cycle stage determines the required information in extraction process. Mapping (and possibly transforming) large-scale research metadata to linked data is required to interlink and integrate different types of data.

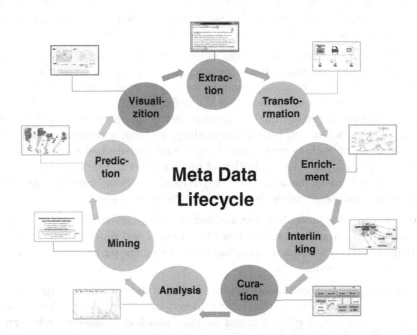

Fig. 2. Conceptual lifecycle for the management of scholarly communication metadata.

Enrichment: (Semi-)automatic and manual semantic enrichment of metadata in knowledge graphs are becoming increasingly popular for information discovery and recommendation services to explore and suggest information about items of interest. Semantic enrichment is a process that assigns quantitative similarity measure by adding statements to the metadata about an object [5]. It can be used to discover related patterns and missing relationships between semantically similar or related items. In consequence, knowledge discovery and ranking services can be provided on top of the graph concepts. The set of identified metadata is semantically enriched by linking and integrating with upper level ontologies. We assign quantitative similarity measures by adding statements to the metadata.

Interlinking: Linked Data interlinking means discovering ideally all instances that represent the same real-world object located in different datasets. To digitize scholarly communication beyond mere open access a comprehensive amount of cross-domain data is required. In order to increase the comprehensiveness of the collected metadata connecting it to relevant data is necessary. We adopt several linking approaches yielding high precision and recall for scholarly metadata interlinking [1].

Curation: This process supports ingestion, semantic lifting and integration of relevant information from various sources. Data curation enables data discovery and retrieval, maintain quality, add value, and provide for re-use over time. We are using two methods of data curation: 1. crowd sourcing methods in which community members are contributing small pieces of information and 2. constant and continuous data curation by the original data providers.

Analysis: Quality metrics are "procedure[s] for measuring a quality dimension", which "rely on quality indicators and calculate an assessment score from these indicators using a scoring function" [3]. The quality of assessed objects is analysed during this stage based on the defined metrics. This stage is about implementations of certain quality metrics specific to the domain of Scholarly Communication which can be used for ranking, filtering and recommending different component such as events in a flexible and user-defined way. The resulting framework supports the definition of quality aspects which are relevant for different stakeholders including authors/researchers, event participants, event organizers, publishers, reviewers, sponsors and organizations.

Mining: A lot of information is embedded inside the scientific articles as static content in the form of fixed successions of characters and words. Mining methods are required to extract such information until having the paradigm change of the way scholarly communication.

Prediction: Due to the dynamic nature of science, the knowledge graphs related to scientific communication can be considered incomplete by default because relations among graph entities might be unestablished, unknown or broken at the time of graph creation. In addition, many of the ranking and quality criteria in the context of scholarly communication are about the *impact* of that particular object on research. Basically, assessing the impact of something recent will require looking into the future. The knowledge graph of such a metadata management system can be used to offer predictions about what status will any object obtain in the future such as citation impact or topic movement.

Visualization: Metadata of scholarly communication is heterogeneous with a large variety of entities and many types of interrelationships. Visualization of such a heterogeneous metadata graph using different views and models such as timelines, calendars, and etc. can help users to have a better understanding. As a final stage of the metadata life-cycle, metadata visualization is required.

In this thesis, we focus in particular on some of these steps: metadata identification, transformation, enrichment, interlinking, curation, analysis and visualization. The other stages are planned to be considered systematically as future work.

4 Methodology

We follow different methodologies in the development of this doctoral work adheres the following tasks: (a) Survey of the state-of-the-art, including semantic enrichment and quality assessment approaches published in the Semantic Web and recommender systems areas. (b) Definition of the problem and solution, and characterization of the properties of the proposed approach. (c) Empirical evaluation of the proposed solution to measure its performance with respect to state-of-the-art approaches. The experiments will be conducted execution of experiments and statistical studies of the obtained results to draw conclusions about the hypotheses and analysis of the results.

5 Results

Different elements of the proposed approach are applied in several related projects. The metadata **identification** and **enrichment** as well as **quality analysis** has been applied in the domain of OpenCourseWare [9] and scientific events [12]. We have **transfered** a recent snapshot of the *OpenAIRE*[8] data to RDF [10]. The OA-LOD is accessible as a complete RDF dump as well as a SPARQL endpoint[9]. **Interlinking** of the OA-LOD and DBLP[10] provides complex **query execution and analysis** of multiple and independent metadata domains of scholarly communication such as relations between Projects, Publications, Fundings, Datasets.

OpenResearch[11] as a collaborative management system for scholarly communication metadata is supporting various metadata life-cycle stages that are considered in this thesis [11]. The metadata **extraction** and **curation** is done using a crowd-sourcing approach. Quality based analysis of metadata are available through form-based, pre-defined and user-defined query executions. Results

---

[8] https://openaire.eu/.
[9] http://beta.lod.openaire.eu/.
[10] https://datahub.io/de/dataset/l3s-dblp.
[11] http://openresearch.org/.

of the assessments are represented using **visualizations** plug-ins such as map, timeline and calendar views. The knowledge graph is available as RDF dump and SPARQL endpoint. The following queries demonstrate that the system is able to answer questions which are otherwise cumbersome or impossible to answer for researchers. **Q.1** List upcoming events related to semantic web which have an acceptance rate lower than 25% will take place in Europe?

```
SELECT ?event ?start ?end ?acceptanceRate ?continent
WHERE {
  ?e rdfs:label ?event.
  ?partContinent rdfs:subClassOf ?continent.
  ?continent rdfs:isDefinedBy site:Category:Europe.
  ?e a category:Semantic_Web.
  ?e property:Acceptance_rate ?acceptanceRate. }
  FILTER(?acceptanceRate<25.0 && ?start>="2016-01-01"^^xsd:date
                      && ?end < "2017-01-01"^^xsd:date)
  ORDER BY DESC(?start) LIMIT 100
```

Listing 1. Upcoming high profiled events wrt. location and time

Q.2 List possible PC members that have been involved in semantic web related events in the last three years?

```
SELECT ?event ?start ?end ?pcMember ?geMember
WHERE {
  ?e rdfs:label ?event.
  ?e property:Has_PC_member ?pcMember.
  ?e a category:Semantic_Web.
  ?e property:Start_date ?start.
  ?e property:End_date ?end.
  MINUS { ?e property:Has_PC_member urir:Some_person. }
  FILTER(?start>="2010-01-01"^^xsd:date &&
       ?end<"2017-01-01"^^xsd:date).
} ORDER BY DESC(?start) LIMIT 10
```

Listing 2. PC member recommendation

The generated knowledge graph is used in a collaborative authoring platform developed by the *OSCOSS* project[12] for **recommendation** services for different stakeholders of scientific communication. The annotated keywords from the content of the under-production scholarly document is used as matching patterns in the graph and reflects recommendations based on similarity metrics.

6   Conclusions and Future Work

This thesis contributes to the fields of information science and scientific communication. The analysis of the state-of-the-art results in a *terminology* and

---

[12] http://eis.iai.uni-bonn.de/Projects/OSCOSS.html.

a *classification* that serves as reference for the comparison and discussion of services and systems facilitating scholarly communication. Based on the conceptual framework of scholarly communication metadata domains and the lifecycle model new services were developed for facilitating the metadata management and thus the digitization of scholarly communication in general.

Acknowledgements. This thesis is supervised by Sören Auer and Christoph Lange at the EIS group at University of Bonn. I would like to thank to Maria-Esther Vidal and Rainer Manthey for their guidance and fruitful discussions. This work has been partially funded by the European Commission for the project OpenAIRE (GA no. 643410).

References

1. Alexiou, G. et al.: OpenAIRE LOD services: scholarly communication data as linked data. In: SAVE-SD Workshop at WWW Conference, SAVE-SD (2016)
2. Auer, S., Bryl, V., Tramp, S. (eds.): Linked Open Data – Creating Knowledge Out of Interlinked Data. LNCS, vol. 8661. Springer, Cham (2014)
3. Bizer, C., Cyganiak, R.: Quality-driven information filtering using the WIQA policy framework. J. Web Sem. **7**(1) (2009)
4. Bryl, V., et al.: What's in the proceedings? Combining publisher's and researcher's perspectives. In: SePublica (2014)
5. Charles, V., Stiller, J.: Evaluation of metadata enrichment practices in digital libraries: steps towards better data enrichments. In: SWIB (2015)
6. Iorio, A.D., Lange, C., Dimou, A., Vahdati, S.: Semantic publishing challenge – assessing the quality of scientific output by information extraction and interlinking. In: Gandon, F., Cabrio, E., Stankovic, M., Zimmermann, A. (eds.) SemWebEval 2015. CCIS, vol. 548, pp. 65–80. Springer, Cham (2015). doi:10.1007/978-3-319-25518-7_6
7. Lange, C., Iorio, A.: Semantic publishing challenge – assessing the quality of scientific output. In: Presutti, V., Stankovic, M., Cambria, E., Cantador, I., Iorio, A., Noia, T., Lange, C., Reforgiato Recupero, D., Tordai, A. (eds.) SemWebEval 2014. CCIS, vol. 475, pp. 61–76. Springer, Cham (2014). doi:10.1007/978-3-319-12024-9_8
8. Pedrinaci, C., Cardoso, J., Leidig, T.: Linked USDL: a vocabulary for web-scale service trading. In: Presutti, V., d'Amato, C., Gandon, F., d'Aquin, M., Staab, S., Tordai, A. (eds.) ESWC 2014. LNCS, vol. 8465, pp. 68–82. Springer, Cham (2014). doi:10.1007/978-3-319-07443-6_6
9. Vahdati, S., Lange, C., Auer, S.: OpenCourseWare observatory – does the quality of OpenCourseWare live up to its promise? In: LAK. ACM (2015)
10. Vahdati, S., Karim, F., Huang, J.-Y., Lange, C.: Mapping large scale research metadata to linked data: a performance comparison of HBase, CSV and XML. In: Garoufallou, E., Hartley, R.J., Gaitanou, P. (eds.) MTSR 2015. CCIS, vol. 544, pp. 261–273. Springer, Cham (2015). doi:10.1007/978-3-319-24129-6_23
11. Vahdati, S., Arndt, N., Auer, S., Lange, C.: OpenResearch: collaborative management of scholarly communication metadata. In: Blomqvist, E., Ciancarini, P., Poggi, F., Vitali, F. (eds.) EKAW 2016. LNCS, vol. 10024, pp. 778–793. Springer, Cham (2016). doi:10.1007/978-3-319-49004-5_50
12. Vahdati, S., et al.: Towards a comprehensive quality model for scientific events and publications. Technical report (2016)

Author Index

Printed in the United States
By Bookmasters